Linden Baier 1

MW00462895

EMERGENT ECOLOGIES

EMERGENT ECOLOGIES

. . .

Eben Kirksey

DUKE UNIVERSITY PRESS

Durham and London

2015

© 2015 Duke University Press
All rights reserved
Printed in the United States of America
on acid-free paper ∞
Designed by Amy Ruth Buchanan
Typeset in Chaparral Pro and Trade Gothic by Copperline

Library of Congress Cataloging-in-Publication Data
Kirksey, Eben, [date] author.
Emergent ecologies / Eben Kirksey.
pages cm
Includes bibliographical references and index.
ISBN 978-0-8223-6017-9 (hardcover : alk. paper)
ISBN 978-0-8223-6035-3 (pbk. : alk. paper)
ISBN 978-0-8223-7480-0 (e-book)
1. Conservation biology. 2. Ecological assessment
(Biology) 3. Ecosystem management. I. Title.
QH75.K58 2015
333.95'16—dc23
2015019795

Cover art: Communications towers in the cloud forest of
Monteverde, Costa Rica. Photograph by Eben Kirksey.

DUKE UNIVERSITY PRESS GRATEFULLY ACKNOWLEDGES THE
SUPPORT OF THE AUSTRALIAN RESEARCH COUNCIL, WHICH
PROVIDED FUNDS TOWARD THE PUBLICATION OF THIS BOOK.

KATE, THIS ONE'S FOR YOU. HI! BIG THUMBS UP.

CONTENTS

INTRODUCTION

. . .

Emergent dynamics can destroy the existing order. Microbes that become emergent diseases—by finding novel exploits, pathways of transmission, or modes of existence—can quickly transform dominant political strategies, economic systems, or agricultural practices.[1] Emergences can also figure into collective hopes.[2] When a forest is clear-cut by loggers or destroyed by a volcanic eruption, emergent plants are the first to sprout. Nascent associations are able to exploit faults and fissures within established assemblages. They contain the promise of supplanting deeply rooted structures. Materializing in interstitial spaces, between divided forces, emergent forms of life can disrupt ostensibly unified systems. False starts in one direction can become significant beginnings along a new vector. Flying in the face of long-term agendas, unexpected detours and happy accidents can generate a novel sense of order.[3]

Emergent Ecologies is a study of multispecies communities that have been formed and transformed by chance encounters, historical accidents, and parasitic invasions. Insights from contemporary philosophy are used to reframe key problems in the field of conservation biology—relating to invasive species, extinctions, environmental management, and reforestation. Following the flight of capital and the trajectories of multiple species across national borders and through fragmented landscapes of the American tropics—from Panama to Costa Rica to the United States and back again—this book asks: How do certain plants, animals, and fungi move among worlds, navigate shifting circumstances, and find emergent opportunities? When do new species add value to ecological associations, and when do they become irredeemably destructive? When should we let unruly forms of life run wild, and when should we intervene? Instead of regarding the past as a legacy that should always be restored, this book focuses critical attention on present interests in ecological communities as well as their possible futures.

Do ecosystems exist in the world? Are they figments of the mind? If destroyed, will multispecies communities predictably reemerge? The roots of these questions go back to a contentious debate between two early twentieth-century biologists: Clements and Gleason. Frederic Clements, who led the botany department at the University of Minnesota, understood ecological associations as natural units of vegetation. A 1916 monograph by Clements described ecological units, like rain forests, marshes, or riparian woodlands, as "complex organisms." These superorganisms, according to Clements, involve stable associations of plants and animals. Following major ecological disturbances and destruction, he found some evidence that these complex associations would come back. Henry Gleason, of the New York Botanical Garden, published a paper in 1926 challenging the influential ideas championed by Clements. Gleason understood ecological associations as relationships in constant flux, arguing that they should not be understood as "an organism, scarcely even a vegetational unit, but merely a coincidence." According to Gleason, ecological communities are not part of the natural order of things, but instead are bounded by artificial lines that reflected the tendency of the human species "to crystallize and classify [our] knowledge."[4]

A. G. Tansley, who coined the term "ecosystem" in 1935, made arguments allied with Gleason: "The systems we isolate mentally are not only included as parts of larger ones," he wrote, "but they also overlap, interlock and interact with one another." While Tansley himself assumed that these systems were in constant flux, many contemporary ecologists have made his idea of the ecosystem unnecessarily concrete.[5] In 1981 Paul and Anne Ehrlich compared ecosystems to airplanes. They argued that it would be terrifying to ride on a partially disassembled flying machine: "As you walk from the terminal toward your airliner, you notice a man on a ladder busily prying rivets out of its wing. Somewhat concerned, you saunter over to the rivet popper and ask him just what the hell he's doing." Ehrlich and Ehrlich think that we should be terrified to live in ecosystems where essential parts, species, are being driven extinct—being popped out of finely tuned systems like rivets.[6]

Popular metaphors are being questioned as a new generation of biologists are describing the emergence of what they term "novel ecosystems." Joseph Mascaro, a plant biologist, rejects the airplane comparison, writing, "Ecosystem function does not solely reflect species loss, as implied by the popping of rivets, it also reflects species additions."[7] Novel ecosystems "are diverse but invaded, neglected but resilient, anthropogenic but wild," in the words of Laurie Yung and colleagues.[8] Ecologists are starting to look for intellectual allies in studying the social and political forces at play within these "messy and neglected wrecks." They are developing approaches to conservation that encourage people to engage with forms of life that exist all around us, abandoning previous efforts to preserve visions of pristine nature. Other biologists have expressed reservations about this conceptual shift: "In today's predominant consumer culture there is a social value that ascribes worth to novelty," write Rachel Standish and colleagues. "The concern, then, is that people will value novel ecosystems simply because they are new."[9]

Lately ecosystems have been shaped by competing ethical, political, and economic values. But the underlying dynamics at work in ecosystems are not necessarily "novel." Researchers in the field of geology and paleontology have recently given Henry Gleason (of the Clements-Gleason controversy) "a gift reserved for but a few theoreticians: irrefutable proof." Tree species have moved "as individuals and not as part of discrete communities or organisms" during times of environmental change in the relatively recent past (from 8,000 to 14,000 years ago).[10] In other words, the rivets that theoretically underpin ecosystems are often moving around on geological time scales, disappearing and reappearing, in a given locale.

Ecosystems have long been shaped by the loss of previous species, the acquisition of new organisms, and the emergence of novel multispecies assemblages.[11] Following Donna Haraway's "Cyborg Manifesto," this book takes "pleasure in the confusion of boundaries" at the margins of ecosystems and makes arguments "for responsibility in their construction."[12] Departing from anachronistic depictions of past environments, I consider the intersecting forces that shape present multispecies communities, as well as possible futures. *Emergent Ecologies* chronicles the actions of people whose instrumental use of certain critters, or love for some kinds of life, has led them to construct novel ecosystems—bringing machines, industrial supply chains, and biological elements together into unusual assemblages. Other forces and agents of assembly—diverse animals, plants, and fungi with their own interests and desires—are also at work in ecosystems emerging around us. People and other beings are becoming entangled in what Isabelle Stengers calls relations of reciprocal capture.[13]

Beings who fold one another into the enduring relationships of reciprocal capture, according to Stengers, often reach symbiotic agreements.[14] Transformative encounters, seductive moments that generate new entangled modes of coexistence, take place when two beings capture one another in a reciprocal embrace. Symbiosis, in the eloquent prose of Lynn Margulis and Dorion Sagan, involves "the co-opting of strangers, the involvement and infolding of others."[15] Symbiotic associations involve beings with a mutual interest in the continued existence of one another.[16] Symbiotic attachments, in Stengers's mind, are not categorically different from other forms of reciprocal capture—like parasite-host entanglements or predator-prey relations, where one party to the relationship is constantly trying to escape, evade, or destroy the other. The visual and cognitive abilities of the bird are brought into being by the camouflage of the caterpillar, which make it difficult to discern against a backdrop of foliage. The host's immune system, odor, and skin refer to the existence of the parasite and its clever modes of detecting the host and invading its body.[17] Beings are coinvented in relationships of reciprocal capture; they "integrate a reference to the other for their own benefit," forming a shared milieu, an environment.[18]

Parasites are key players in emergent ecologies. The word "parasite" is polysemic in French—meaning biological or social freeloader in addition to "noise" or "static."[19] Michel Serres celebrates the productive and creative nature of noise in his playful monograph, *The Parasite*. Parasites are jokers or wild cards, Serres claims, who take on different values depending on their positions. "The parasite doesn't stop," writes Serres. "It doesn't stop eating or drinking or yelling or burping or making thousands of noises or filling space with its swarming and din. . . . It runs and grows. It invades and occupies."[20] Within the realm of tropical ecology, parasites and pathogens are regarded as forces that generate diversity. The Janzen-Connell hypothesis, a widely accepted explanation for tree species biodiversity in tropical forests, suggests that specialized insect herbivores, bacteria, viruses, and fungi reduce the numbers of common trees. Seedlings that germinate farthest from their parents should have an advantage since they are far from the species-specific parasites and diseases targeting other members of their kind.[21]

Emergent Ecologies describes parasitic invasions that destroyed established communities while simultaneously opening up new possibilities for flourishing.[22] A microscopic fungal disease that has pushed thousands of frog species to the brink of extinction is a central figure in my entangled tales.[23] Diverse technological apparatuses, scientific enterprises, market economies, and forms of life have been brought together to save frogs from this fungus. While describing the artificial ecosystems that have been constructed around literal amphibians, this book also explores the lifeways of "ontological amphibians"—insects, varieties of rice, and monkeys that are constantly moving among worlds, deciding which ontology they would like to inhabit.[24] Alongside endangered forms of life, I found a swarming multitude that was constantly creating new symbiotic associations, taking advantage of exploits in emergent ecosystems, and going wild along unexpected trajectories.[25]

Wild creatures are often understood as having an "existential independence" from human worlds.[26] Rather than treating wildness as a phenomenon that exists only beyond the reach of civilization or domestication, this book also focuses on the risky and out-of-control dynamics that emerge amid intimate entanglements with other species.[27] Contagious excitement and fear often accompany moments of capture, when humans involve and enfold other creatures into a new association. Mixed emotions are also at play when we release others from our care, allowing them to escape our tentative grasp. While some cultural critics have characterized conservationists as "misanthropes," as melancholics who see humans as inherently destructive while regarding other species as essentially good and innocent, my aim is to offer a more nuanced characterization of the desires, affective attachments, and dreams motivating people to care for wild things and living systems.[28]

Novel ecological assemblages are being created by expert practitioners, as well as by amateurs embracing a Do-It-Yourself (DIY) ethos, people who are experimenting with new ways of living responsibly with other critters in multispecies worlds.[29] Human interactions with animals have driven recent ethical debates in anthropology, history, and contemporary philosophy.[30] Departing from "the question of the animal," the polemic by Jacques Derrida arguing that "the human-animal distinction can no longer and ought no longer be maintained," *Emergent Ecologies* also engages with "the question of the fungus" and "the question of the plant."[31] Fungi illustrate "practices that thrive in the 'gap' between what is taken as wild and what is taken as domesticated," according to the Matsutake Worlds Research Group. "Thinking like a fungus" opens up questions like, Who is doing the domesticating? And to what end? *Plant Thinking*, by Michael Marder, regards plants as "collective beings," as "non-totalizing assemblages of multiplicities, inherently political spaces of conviviality."[32] Other beings who have "strivings, purposes, telos, intentions, functions, and significance" come together in Eduardo Kohn's book, *How Forests Think*.[33]

Following plants, animals, and microscopic fungi as they became caught in temporary entanglements, and then escaped, *Emergent Ecologies* uses the methods and tactics of multispecies ethnography to trace the contingencies of unexpected connections.[34] Conventional ethnographic interviews with biological scientists, environmental activists, and others living in the shadows of conservation initiatives were supplemented with original historical research in archival collections, my own biological experiments, and artistic interventions. Artists who cleverly use scientific equipment with a DIY ethos—to track the flight of pigeons in polluted urban air, or to listen to the laughter of laboratory rats—have inspired many ethnographers to adopt new tactics and techniques for studying biological subjects.[35] Venturing into the realm of microscopy as a participant observer, I noted the presence of beings and things at the periphery of the scientific imagination. Investigating the shared worlds of humans and animals led me to borrow methods from the field of ethology, a discipline based on the direct observation of animal behavior. Ethological methods have long been wedded to explanatory frameworks focused on either proximate mechanisms or ultimate (evolutionary) functions.[36] Departing from conventional ethological techniques, which record and quantify predicted behaviors, I employed flexible and open-ended descriptive techniques for noting and filming behaviors in multispecies worlds.

Material gathered from diverse sources forms the basis of my interlocking tales from multiple sites in the Americas—from the Canal Zone of Panama to art galleries of New York City, riparian woodlands of Florida, and abandoned pasturelands of Costa Rica. These tales all speak to key questions: Which creatures are flourishing, and which are failing, at the intersection of divided forces, competing political projects, and diverse market economies? Amid widespread environmental destruction, with radical changes taking place in ecosystems throughout the Americas, where can we find hope? Holding onto hopes for the continued existence of vulnerable beings, like members of an endangered species, risks the possibility of cruel disappointment if they do indeed disappear. Even still, the maintenance work required to enhance the flourishing, endurance, and survival of critters in a precarious condition is more necessary than ever.[37]

Contemporary writing on the environment is largely focused on doomsday scenarios. *Emergent Ecologies* departs from this dominant plotline, insisting that we reject apocalyptic thinking.[38] Against the backdrop of pervasive fears, this book explores the possibility of grounding hopes in shared futures. Living with contingencies in shared worlds, navigating circumstances and forces beyond our control, requires imaginative as well as practical labor. Rather than remaining anxiously focused on possible losses, this book explores the imaginative horizons of organic intellectuals who are sifting through the wreckage of catastrophic disasters, searching for hope within landscapes that have been blasted by capitalism and militarism. Reaching into the future, these thinkers and tinkerers are grabbing on to hopeful figures and bringing them into existence in the present.[39] Tactfully guiding interspecies collaborations, new generations are learning how to care for emergent assemblages by seeding them, nurturing them, protecting them, and ultimately letting go.[40]

CHAPTER ONE

PARALLAX

. . .

Barro Colorado Island is an "open-air biological laboratory" in the Panama Canal. It is run by the Smithsonian Tropical Research Institute, and its activities are outwardly united by a single goal: "to increase understanding of the past, present, and future of tropical biodiversity and its relevance to human welfare."[1] At this facility, long-term projects have been established around specific research questions and focal organisms: the dispersal of seeds by mammals, the population dynamics of canopy trees, the pollination ecology of euglossine bees, symbiosis and parasitism in fig trees, and the neuroethology of bats. Many biologists on Barro Colorado Island are knowledgeable about the historical forces that have shaped their island laboratory. Egbert Leigh's influential textbook, *Tropical Forest Ecology: A View from Barro Colorado Island*, clearly states, "Barro Colorado's biota is in no sense 'pristine.'"[2] Despite this historical consciousness, I found depictions of reified Nature in the Smithsonian's archives that only gave room for some forms of culture.

Rewinding past more than one hundred years of history, to early U.S. military adventures in Central America, and then fast-forwarding back again, produces a parallax effect—a mode of three-dimensional depth perception that emerges when nearby objects move against a distant backdrop. Early visitors who toured the Panama Canal experienced "stereoscopic visions," in the words of Ellen Strain, where tourism doubled as a mode of time travel. Learning to view the landscape through hand-operated stereoscopes, containing a pair of photographs that used the parallax effect to produce three-dimensional illusions, visitors came to view Panama "as the ideal tourist object with its natural wonders—tropical fruits, luxuriant vegetation, the Rio Grande River, fresh water springs, and scenic bays—and its combination of an intriguing past, an exotic present, and a bustling future which lies ahead."[3] Contemporary scien-

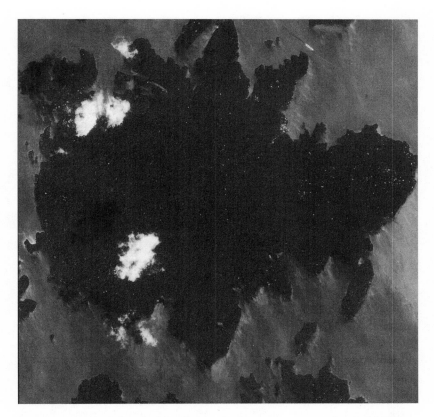

FIGURE 1.1. Barro Colorado Island as seen from outer space by QuickBird, a high-resolution satellite. A massive cargo ship, transiting the canal, can be seen in the top right-hand quadrant of this picture. Image from NASA's Earth Observatory.

tific objects—like euglossine bees, symbionts, and parasites—gain depth when viewed against the backdrop of these earlier objects of wonder and when situated within the political and economic forces that shaped the ecology of Central America. President Teddy Roosevelt helped create the nation of Panama in 1903, supporting separatist insurgents and initiating a naval blockade against Colombia. On the heels of this military action, the United States took over the construction of the Panama Canal—a spectacular marvel of engineering that facilitated the flow of global commerce, fortified an emergent empire, and created a "living laboratory."[4]

Barro Colorado Island was gradually created by the rising waters of the Chagres River after U.S. engineers installed a dam in 1914 during the construction of the Panama Canal. Widespread cutting and clearing of the forest from the nineteenth century, from the French attempt to build a canal, left a lasting impact on the northeastern half of the island.[5] After Barro Colorado Island was declared a reserve in 1923, an ecosystem emerged that served U.S. strategic interests and the desires of biological scientists. Seven Panamanian farmers who remained on the island were eventually forced to leave. "Remnants of the plantings of bananas, oranges, limes, guava, etc., are still encountered in the bush," according to a 1924 clipping in the Smithsonian archives, "although all cultivation by natives is now a thing of the past."[6] "The island is set aside solely for the purposes of scientific study," according to another early archival document, "and hence no hunting permits, or tree-cutting permits will be issued for this natural preserve except for scientific purposes."[7]

Administrators were preoccupied by Panamanian incursions onto this nature reserve even though activities by the U.S. government were arguably much more ecologically destructive. The Smithsonian archives are relatively silent on the impacts of the Panama Canal on local flora and fauna, as some twenty-two square miles of rain forest and farmlands were expropriated by the United States and drowned under floodwaters.[8] Hunters with historical ties to the lands of the Canal Zone became "poachers" who were routinely fined, detained, and sometimes assailed with gunfire. While researchers from the United States were given permits to capture, kill, and collect animals and plants on Barro Colorado Island, other uses of forest resources by Panamanians were strictly prohibited. "Whatever destruction takes place is that which is ordered by Nature and which is the law of the wild," according to a 1931 article by James Zetek, the first director of Barro Colorado Island. "Plants grow, reach maturity and die. They have their enemies. So also animals grow and die, must fight for their existence. They also have natural foes. But Man is out of this picture. When he comes to the island he is a peaceful intruder. He comes to study, not to destroy."[9]

James Zetek's own research on Barro Colorado Island was focused on the destruction of certain kinds of life. Financial backing from the American Wood Preserver's Association, the Grasselli Chemical Company, and the Southern Pine Association, among other sources, enabled him to test the efficacy of a variety of poisons on termites.[10] Alongside research initiatives that directly aided U.S. commerce, a multitude of other projects emerged within the architecture of empire. Despite being dramatically shaped by industry and agriculture, this man-made island quickly became the premiere site in the Americas for studying tropical ecology. The island was viewed by early researchers as an exotic field site for adventures in the present, which contained the mysterious secrets of nature's past, where new discoveries might unlock future possibilities. It became a site of pilgrimage for aspiring scientists. Visiting became "a rite of passage," in the words of Pamela Henson, a historian at the Smithsonian Institution Archives. "A field trip to the tropics [was] a route to fame for young North American naturalists."[11]

The historical archives of the Smithsonian are full of accounts by young men whose lives and careers were transformed by encounters with other forms of life in the Panama Canal Zone. Surprising behaviors by monkeys and ants, as well as uncanny features of plants and fungi, captured the imagination of visiting researchers and prompted new studies of ecological interdependency.[12] As Barro Colorado Island became a key institution supporting the fledgling discipline of ecology, certain categories of people were excluded from the social world of this new science. Social separation was naturalized among humans even as ecological entanglements were discovered. An architecture of apartheid initially separated men from women and whites from "coloreds" at this Smithsonian research station.[13] "The first women to conduct field work in the tropics encountered many of the well-known barriers to professional women," writes Pamela Henson, "as well as the challenges of dealing with unfamiliar environments and cultures."[14] Disputes about whether or not groundskeepers of "white descent" should have the privilege of using the white toilet, the same toilet used by researchers, were among the contentious subjects animating the correspondence among founders of the biological station.[15]

Barro Colorado Island, with its sharply divided social worlds, was a microcosm of the Canal Zone—a place of U.S. military operations that was off-limits to Panamanian citizens who did not carry a special pass.[16] Gamboa, the nearest town, was designed by the U.S. government "to reflect and facilitate a system of industrial relations based on a rigid class and racial hierarchy . . . with a sharply segregated workforce divided by a dual wage system into 'gold' (white/U.S.) and 'silver' (non-white/non-U.S.)." The Smithsonian Tropical Research Institute began administering Barro Colorado Island in 1946. Even after the dual wage system was abolished in 1948, segregation continued to be "a powerful institutional and cultural force" in the Canal Zone. The architecture of many buildings, such as the clinic, contained separate entrances, waiting rooms, examination rooms, physicians' offices, and overnight quarters for "silver" and "gold" social categories.[17]

Those who were privileged enough to belong to the gold social category participated in a government-sponsored utopia. The gold workers enjoyed a stable and comfortable lifestyle in the policed atmosphere of the Canal Zone—with sports facilities, movie theaters, and churches all built in a series of planned towns. Social harmony and stability, very much in line with Sir Thomas More's original novel *Utopia* (1516), prevailed there in contrast to the chaotic state of affairs in Panama and Central America more broadly. Like other utopian projects, there was what Foucault regards as "a panopticon effect" in these planned towns, with spatial systems of surveillance built into the landscape. In Gamboa, workers' houses were carefully arranged, with the lowest-paid living close to the canal in a low valley, and the higher-paid supervisors living on a ridge overlooking the scene. Imagination and authoritarianism came to life there, as part of the broader midcentury new urban movement described by David Harvey in *Spaces of Hope*.[18]

Expatriate U.S. citizens who took up long-term residence within this Canal Zone utopia began calling themselves "Zonians." A third-generation Zonian, who masquerades online under the anonymous username of killbyte, has posted photographs on Flickr and snippets of text that offer candid views of a social world united by doing fun things together amid a military occupation: "I am indeed part of a small, privileged group that belong to a dwindling, elite club that will never exist again. Yes, perhaps it was an experiment in US colonialism—they made sure we retained our US heritage by importing everything cultural that made us feel like US citizens, but we were distinct enough in the sense that we could go into the rain forest & use it as our own private playground. The jungle swimming holes were amazing!"[19]

Forested ecosystems on the banks of the Panama Canal, and the surrounding watershed, became linked to the life of global commerce in the 1970s. Ashley Carse, a cultural anthropologist, suggests that nature became part of the U.S. government "infrastructure" in Panama for storing water and regulating its flows. Agricultural methods of Panamanian farmers, which involved periodically clearing the forest in swidden systems, came to be seen as "the specter of commercial death" for the canal. The forest for these campesinos was not a fixed object, a green space on the map, but a dynamic system, an emergent ecology, tied to their own economic livelihoods. A 1978 essay by Frank Wadsworth, "Deforestation: Death to the Panama Canal," mentions a number of factors contributing to water scarcity in the canal system—drought, ship traffic, and municipal water use. But, ultimately, only Panamanian agricultural practices were targeted by policy makers. Parklands and nature monuments were created as farmers were pushed from their lands with a combination of financial incentives and military operations.[20] Policing operations in the forest where expatriate Smithsonian scientists worked also intensified. "Poachers still roamed the more distant portions of Barro Colorado almost at will in the early 1970s," writes Egbert Leigh in *Tropical Forest Ecology*, "but poaching on the island was almost entirely suppressed by 1985."[21]

By 1997, when I made my own initial pilgrimage to Barro Colorado Island as an undergraduate research assistant, armed forest rangers (*guardabosques*) were still a visible presence at Smithsonian facilities. The entitlements of white Zonians were rapidly dwindling, even though some measures of distinction and segregation were in place. My U.S. passport continued to grant me privileges—like entry to the old Officer's Club on Clayton Army Base. My citizenship also facilitated my initial access to Smithsonian facilities. A forty-minute boat ride separated the Smithsonian's living laboratory from Gamboa, and it remained inaccessible to ordinary Panamanians who could not afford to pay for a day-long guided nature tour. In the 1990s, local historical memories were haunted by the 1989 U.S. invasion of Panama that killed some three thousand civilians and deposed President Manuel Noriega (who had formerly been regarded as a CIA "asset").[22] Future uncertainties also loomed large on the horizon. The United States was slated, in accordance with international treaties, to give the Canal Zone to the nation of Panama on December 31, 1999. But messages from powerful political factions in Washington signaled that the planned transfer of sovereignty might not take place.

The project that initially brought me to Panama as an undergraduate assistant was indirectly in the service of U.S. geostrategic interests—it was research that would potentially benefit the citrus industry. The electric ant (*Wasmannia auropunctata*), an insect native to Panama, had become a common agricultural pest in the southern United States. Fruit pickers were demanding premium wages to work in infested orange groves in Florida, because the ants can deliver a painful sting, like an electric shock. They were notorious for swarming inside the workers' clothes. The electric ant had become a cosmopolitan insect, ranging over many different countries, free from national limitations or attachments.[23] Spreading in areas disturbed by humans, it had invaded ecosystems emerging around human agricultural schemes. Hitching a ride in shipments of produce, nesting in rolled leaves or dead sticks or almost anywhere, this nomadic species had taken up residence in West Africa, Melanesia, Polynesia, and islands throughout the tropical Americas.[24]

Electric ants in Panama live within a diverse community of other ants

FIGURE 1.2. *Ectatomma* is a genus of ant found throughout Central and South America comprising many different species. My own encounters have mostly been with *E. ruidum*, one of the most common ants in the forests of Panama and Costa Rica. This species lives in groups of around 150–300 individuals in underground nest chambers with pupae and larvae. The unadorned holes of *E. ruidum* colonies are easy to distinguish from those of *E. tuberculatum*, a slightly bigger ant that builds tubular nest entrances at the base of small trees and vines. This picture of *E. tuberculatum* was taken in Gamboa, Panama. Photograph courtesy of Alex Wild. (Alex Wild has an outstanding photographic portfolio, featuring ants and other insects, available online: http://www.alexanderwild.com.)

in leaf litter on the forest floor. The project that brought me to Barro Colorado Island in 1997 sought to understand if competition with these other species helped regulate electric ant populations. During the fieldwork stage of the project, my role involved placing tuna fish baits at marked spots on the ground, collecting ants at the baits, and identifying them under the microscope back in the lab. While gathering data in tangles of underbrush, dripping with sweat from the sweltering heat, I became familiar with the habits of *Ectatomma ruidum*—one of the largest ants at the baits, which frequently wrestled large chunks of tuna fish away from smaller competitors. I came to easily recognize *Ectatomma* with my naked eye and began to follow these charismatic insects away from the tuna baits, on alternate lines of flight.

Most ant species vigorously defend the boundaries of their colony—killing intruders from different colonies of the same species on contact. For most ant species, the stranger is the enemy "with whom there is the real possibility of a violent struggle to the death."[25] Casual observations of *Ectatomma ruidum* suggested that this species is different from most ants—in a certain sense it is exceptional, in fact. While studying "competition" among leaf litter ants, I found surprising forms of collaboration among *Ectatomma* ants. Spending hours casually watching different colonies, I watched ants carry food, larvae, other workers, and even winged queens between distinct nests. Making my own informal experiment, I put up a barrier around one focal *Ectatomma* colony and let the ants continuously collect tuna fish bait for an hour. After removing the barrier, and the bait, I watched as tuna fish was redistributed. Ants exited the focal colony and carried it into the nests of neighbors. Minutes after watching tuna entering one neighboring nest, I watched as it was carried out again to an even more distant nest. As I spent more time watching *Ectatomma*, I found that guards will sometimes stand in the nest entrance and occasionally bite or drag away *Ectatomma* ants from other colonies that are trying to get inside. But often the nest entrances stand empty. Ants actively guarding the nest entrance also sometimes stand aside, letting members of neighboring nests pass unmolested. Once inside, these ants have access to caches of food.[26]

Marking individual adult ants with paint, and gripping a hind leg with a pair of steel forceps, I positioned them at the entrance of colonies that were not their own. Almost unfailingly, when released, the ants went inside. Conducting my own experimental trials in 1997, I spent close to 150 hours in the field—staring at small holes in the ground, squatting on my knees, waiting for these painted ants to reemerge. In short, during all this waiting and watching I found that *Ectatomma* ants regularly enter the nests of their neighbors. I also discovered that ants from distant nests—more than three hundred meters away—can readily enter the colonies of strangers. Rather than a categorical rejection of all nonkin, I found a nuanced pattern of graded recognition. Hostile acts (low-level biting, dragging, but never fatal stinging) were only occasionally directed toward stranger ants.[27]

In the late 1990s, the era when I made these observations, the genetic determinism of E. O. Wilson's sociobiology held sway among myrmecologists, experts who study ants.[28] In the ideal ant colony (at least according to Wilson and his followers) there is a single queen and all of the workers are sisters: nonreproductively viable females.[29] Sociobiologists were asserting that the ant colony "is a superorganism." Nests of ants were "analyzed as a coherent unit and compared with the organism in the design of experiments, with individuals treated as the rough analogues of cells." In an encyclopedic tome published in 1990, simply titled *The Ants*, Bert Hölldobler and E. O. Wilson speculated that "natural selection can produce selfish genes that prescribe unselfishness."[30] As an undergraduate, majoring in cultural anthropology and biology, I became fascinated by behaviors of *Ectatomma ruidum* that did not fit with the prevailing consensus of the 1990s. Conventional models regarded the ant colony as "a hub, or star, or network in which all lines . . . radiate from a central point along fixed lines." I found that *Ectatomma* ants were entangled in something more like a "distributed, or full-matrix, network in which there is no center and all nodes can communicate directly with all others."[31] If ant colonies were to be understood as superorganisms, my findings about workers moving among colonies suggested that the cells were running wild.

CHAPTER TWO

ONTOLOGICAL AMPHIBIANS

. . .

The social worlds of scientists, ecotourists, forest rangers (guardabosques), and farmers overlap at the Smithsonian Tropical Research Institute in Panama. When I returned to Panama as a National Science Foundation postdoctoral fellow in 2008 I found that each of these worlds was united by at least one primary activity—like bird-watching, gathering experimental data, hunting, or going for nature hikes. Amid shifting political and economic forces, as U.S. dominion was waning and Panamanians were reasserting sovereign powers, I found evidence of negotiation and compromise in the context of seemingly intractable conflicts. But I became drawn away from these dynamics in human social worlds. Dwelling as a participant observer in the Smithsonian's Insect Cognition Laboratory—a crumbling building near abandoned Panama Canal Company barracks and the remnants of U.S. military bases—I was pulled back into the rich social and environmental worlds inhabited by my favorite ant, *Ectatomma ruidum*.

Ectatomma ants were flourishing in emergent ecologies. I found them foraging for insects at night under electric lights, gleaning sugary liquid honeydew from leafhoppers, and communicating with caterpillars in high-pitched stridulatory sounds. In old-growth rain forests, *Ectatomma* were among the most abundant ant species. They were also thriving in closely cropped lawns in the shadows of abandoned U.S. military installations. On the margins of parking lots, and in other zones of abandon, I found these abundant insects multiplying beyond human dreams and schemes. Proliferating within the realm of agricultural and industrial enterprises, within environmental management regimes designed with the well-being of other species in mind, these insects had also become tenacious parasites. They were constantly moving among multiple elements, never just sticking to one environment. In a phrase, I found that *Ectatomma* ants had become ontological amphibians.

This phrase, "ontological amphibian," originates with Peter Sloterdijk, who (following Martin Heidegger) suggests that animals "move around in an ontological cage." Sloterdijk maintains that humans are constantly "switching from one element to another." He claims that our species, *Homo sapiens*, is exceptional: "The human being is a moving animal which longs to change elements and to go somewhere else." Literal amphibians can choose among modes of existence—they can live on earth or in water. Ontological amphibians, according to Sloterdijk, never stick to just one world but always face a decision about what kind of ontology to inhabit.[1] Since Sloterdijk enjoys wide influence, it is worth engaging with his argument on its own terms.[2] His words are also fun to play with. Rather than staying within the confines of his bubbles of meaning, I will push and poke at Sloterdijk's ideas, bringing them into realms where he did not intend for them to travel.[3]

In making "ontological amphibians" my own, I draw on Isabelle Stengers's understanding of cosmopolitics. Stengers offers an idiom for considering the diverging values and obligations that structure possible nonhierarchical modes of coexistence.[4] "The cosmos refers to the unknown constituted by multiple divergent worlds," she writes, "and to the articulations of which they could eventually be capable."[5] Nomads, such as these amphibious insects, can be dangerous, irredeemably destructive, or tolerant, according to Stengers. The challenge, for Stengers, is to trap nomads, to enfold them in production of cosmopolitical worlds—communities that are formed through contingent *political* articulations against the backdrop of the unknown *cosmos*. Forming cosmopolitical worlds means making high-stakes, and potentially groundless, distinctions between enemies and allies.[6] Building cosmopolitical projects involves working together, tooth and nail, in concert with others.[7]

Tracing actions oriented to the care of beings and things, often across species lines, I consider how *Ectatomma* ants have been enlisted in the production of common worlds, and how they escape. These insects are agents of cosmopolitical assembly, conscious beings who become involved with other creatures through relations of reciprocity, kinship, and accountability. Exploring the fleeting whims of these ants, I also consider sentiments about the distribution of surplus that are beyond rational calculus. Studying the promiscuous liaisons of these ants in multispecies worlds led me to augment the conventional tool kit of the social sciences and humanities with an experimental apparatus fashioned out of everyday objects and laboratory equipment. Venturing into the archives of science, I also found others asserting that *Ectatomma* are exceptional animals with a "highly developed social system."[8]

RINGED WORLDS

Peter Sloterdijk's claims about the amphibious nature of the human, and the "ontological cage" that traps the animal, rest on the notion of "environmental world," or *umwelt*.[9] Jacob von Uexküll, an Estonian biologist, coined the word "umwelt" to refer to the phenomena an organism can perceive and also act upon. The German preposition *um* denotes a ring, an enclosure, a surrounding.[10] Conscious beings, according to von Uexküll, are each enclosed within phenomenological bubbles, worlds of perception and action. "Figuratively speaking," writes von Uexküll, "every animal grasps its object with two arms of a forceps—receptor, and effector."[11] Phenomenological worlds (*welten*) are thus constructed by the tentative grasp of each creature.[12] "We human beings cannot enter directly into the *umwelten* of other creatures," writes Timothy Ingold in a critical reappraisal of von Uexküll's work, "but through close study we may be able to imagine what they are like."[13]

If von Uexküll invited his readers to look at familiar places with nonhuman eyes—the eyes of jackdaws, bears, and moths, among other creatures—the disorienting effect of this imagining became the strongest with his description of the tick.[14] "The whole rich world around the tick," writes von Uexküll, "shrinks and changes into a scanty framework—her *umwelt*." Ticks are blind bloodsuckers that, according to the knowledge of von Uexküll's time, attend to only three cues: sunlight, butyric acid (a component of mammalian sweat), and warmth.[15] The poverty of the tick's world "guarantees the unfailing certainty of her actions." The three perceptual cues of the tick generate three distinct activity patterns: sunlight = crawl up; butyric acid = drop; warm hairy membrane = suck.[16]

In the hands of Martin Heidegger, and his postmodern avatar, Peter Sloterdijk, the impoverished umwelt of the tick became a figure of the "ontological cage" of all animals. Animal behavior, for Heidegger, involves "captivation" (*Benommenheit*) by things as opposed to the unbounded and open orientation of humans. Nonhuman organisms, in the Heideggerian tradition, are only capable of relating to those beings that "disinhibit [*enthemmt*]" their behavior, or initiate their capability in some way.[17] Heidegger rested his claims about the exceptional nature of humans on a triple thesis: "The stone is worldless; the animal is poor in world; man is world-forming."[18] Sloterdijk, adding his own twist, suggests, "Man stands

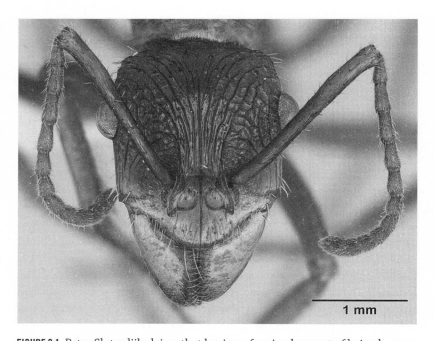

1 mm

FIGURE 2.1. Peter Sloterdijk claims that having a face is a key part of being human. "Human faces have pulled themselves out of their animal form simply by looking at one another," Sloterdijk asserts. The "turning of faces towards other faces among humans became face-creating and face-opening, because the welcome qualities of faces for the eyes of the potential sexual partner inform generic processes via selection-effective preferences" (Sloterdijk, *Bubbles*, 164). Gazing into the eyes of an actual animal, even a creature with very different eyes like *Ectatomma*, I find it difficult to deny that she has a face. Certainly social mammals, who use their faces for all sorts of communication, have particular kinds of faces. Mammal faces are likely unimportant and unintelligible to ants. At the same time, ant faces may create openings toward one another that are difficult for mammals to understand. "The term *interface*," in the words of Juno Parreñas, "implicitly recalls the sense of face developed by Emmanuel Levinas. . . . An ethical obligation to the other is made when one perceives the other's face." If ant faces are frozen in an exoskeleton, lacking muscles to create expressions that might produce empathy in humans, our faces lack antennae—fingeryeyes that enable insects to taste, grope, and smell one another. Being pulled into the distinctive forms of animal faces is an opportunity to consider the radical specificity and limits of the human umwelt in interfaces involving other species (Parreñas, "Producing Affect," 675; Rose, "What If the Angel of History Were a Dog?"). Microscopy and photograph by April Nobile.

open to the world and is indeed in the world (*Welt*) but other animals are caged in their *Umwelt* (ringed-world)."[19]

In contrast to the relatively impoverished umwelt of the tick, the world of *Ectatomma ruidum* is wealthy. This ant has well-developed compound eyes and can remember complex patterns of shadow and light.[20] Like many other insects, *Ectatomma* can see the polarity of light, a dimension of the world to which *Homo sapiens* and most other mammals are blind.[21] In addition to compound eyes, all ants have a pair of antennae, organs that might be called (following Eva Hayward) "fingeryeyes": components of a sensorial ontology, a haptic epistemology, where knowledge comes from touching, tasting, smelling, groping, and reaching.[22] Covered with hairlike sensory organs (*sensillae trichoidea*), the antennae of *Ectatomma* enable them to detect a diversity of chemical compounds, humidity, and heat.[23]

A blog that is all about ants, called Archetype, has scripted a series of embodied exercises to help humans understand what it is like to grasp the world with antennae. While gazing at scanning electron micrographs (figures 2.2 and 2.3), pictures of the sensors on the tip of an ant's antenna, readers might embrace the spirit of Natasha Myers's work. Myers is an ethnographer who has chronicled dances performed by molecular biologists.[24] Following Myers, and instructions from the Archetype ant blog, I invite you to act out this embodied exercise:

> Extend your arms forward with the palms of your hands facing down. Your *sensillae trichoidea* (Latin for hair-like sensory organs) will occur in the greatest number where your thumbs are. This arrangement is particularly suited to smell whatever is in front of your head. . . . in addition to sensing various chemical compounds *sensillae* are involved in sensing humidity and heat. The most common *sensillae trichoidea* covering the hard and otherwise numb exoskeleton of adult ants (and Arthropods in general) are of the mechanoreceptor or tactile type, that is, the sense of touch. If you want to know what it feels like to have an insect sense of touch just gently brush the hairs on your arm.[25]

Some arthropods that use antennae to grasp the world, like ticks, perform a relatively narrow set of behaviors in response to information that they glean from touching, tasting, smelling, and groping objects and other beings. In contrast, social insects, like bees and ants, demonstrate "excellent learning capabilities," in the words of Zhanna Reznikova, a Russian biologist who studies ant behavior. Ants encode complex memories in "mushroom bodies," structures inside their brain that are shaped like

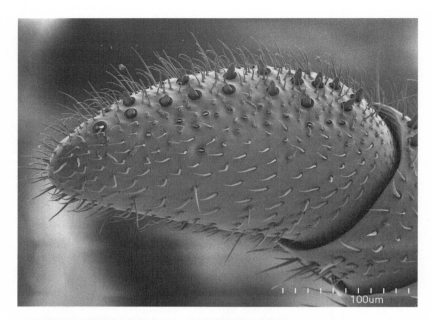

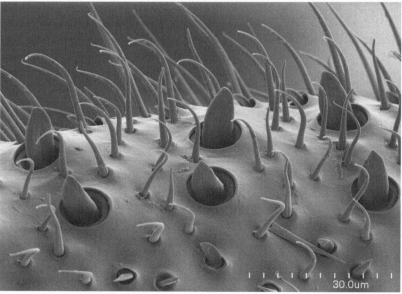

FIGURES 2.2 AND 2.3. Scanning electron micrographs of the tip of the antenna in the African driver ant (*Dorylus helvolus*). The tongue-shaped structures are among the many kinds of *sensillae trichoidea*, the hairlike sensory organs that enable ants to detect heat, humidity, touch, and a diversity of chemical compounds. Microscopy and photographs by Roberto A. Keller/American Museum of Natural History, New York.

fungal caps. "The quantity of neurons does not make the cleverest organism," Reznikova argues. "Memory sits comfortably in mini-brains."[26] Martin Giurfa, a specialist on honeybee cognition, says that biologists once regarded insects as "little robots" or small reflex machines. The latest research suggests that many social insects have plastic, or flexible, behavior. With an ability to master conceptual forms of learning, and nonlinear problem solving, findings suggest that some social insects can categorize, extract, and organize their knowledge.[27]

Ectatomma ants have the capacity for time-place learning, meaning that they associate specific feeding places with different times of day.[28] They communicate with each other by releasing chemical pheromones, through mutual groping and tactile stimulation with antennae, as well as by making chirping noises—produced by rubbing, or stridulating, parts of their exoskeleton. If the umwelt is an ontological cage for some organisms, where stimuli trigger predictable responses, certainly there are species, like *Ectatomma*, that take advantage of surprising encounters. Some species, like *Ectatomma ruidum* and *Homo sapiens*, are more amphibious (in an ontological sense) than other kinds of creatures. Still, we are all enclosed within specific limits of our umwelten. These rings enclose us in phenomenological worlds, structuring the gaps in our gaze toward other creatures across the species interface. Agents locked in reciprocal capture, who inhabit common worlds, can grasp each other—even if they cannot always hold on, even if there are disjunctures in their interests.[29]

INTERESTED OTHERS

Certainly I am not the first human whose imagination has been captured by *Ectatomma*. Dr. O. F. Cook, of the U.S. Department of Agriculture, became infatuated with this insect during a 1902 expedition to the eastern highlands of Guatemala. At that moment the boll weevil, an insect invader from Central America, was beginning to devastate cotton harvests in the United States. Cook discovered that *Ectatomma* ants were attracted to extrafloral nectaries on cotton plants—glands on the leaves, stems, and flowers that secrete a nutritious liquid. Many other species of plants in Central America—like *Inga* trees—also lure ants to their leaves with sugary and nutritious nectar glands. In the case of cotton, these nectaries were inducing ants to patrol cotton crops and kill boll weevils. *Ectatomma* ants have "taken a step toward the domestication of the cotton plant," wrote Cook. "They have at least adopted it, and show an instinctive interest and attraction for it in preference to other plants."[30]

FIGURE 2.4. These two *Ectatomma* foragers have been captivated by a common plant. While waiting for nectar—a sugary and nutritious liquid—these ants help protect the plant from leaf-eating insects. This plant, a species of *Inga*, has captivated some humans too. The flesh of its fruit tastes like vanilla ice cream. These plants (known in English as ice cream bean trees) have enfolded ants and humans together in convivial worlds. Photograph courtesy of Alex Wild.

Imposing the language of late twentieth-century biology on Cook's findings, the relationship between *Ectatomma* ants and the cotton plants was a facultative symbiotic association. Facultative means "optional" or "discretionary." These associations involve contingent and nonnecessary links—in this case meaning that the plant can live without the ant, and vice versa. The Q'eqchi'-speaking Mayans of Guatemala, whom Cook encountered during his expedition, were certainly also aware of this symbiotic relationship. They referred to *Ectatomma* as "the animal of the cotton."[31] Ants, cotton plants, and the Q'eqchi' were entangled in relations of mutual use and exploitation. Caught in a relationship of reciprocal capture, each of these agents had cause to be interested in the continued existence of the others. Clever at co-opting strangers, the ants and plants were both agents of cosmopolitical assembly.[32]

Cook desired to enlist some of these Central American ants in the protection of the cotton plantation economy in North America, an endangered world. Working to recruit allies—humans and multiple other species—he began building an expanding network to stabilize his ideas and proposed interventions. Becoming a gatekeeper for other agents, an obligatory point of passage, Cook was establishing himself as the central node in a project of entrepreneurial *interessement*, to deploy a key word from actor-network theory. "Inter-esse" means being in between or interposed.[33] Like many other scientific entrepreneurs before and since, Cook was interpreting the interests of other species to incorporate them into the dreams and schemes of humans.

The findings from Cook's expedition to Central America were greeted with much fanfare. The *Houston Post* heralded, on its front page, "Enemy of the Boll Weevil: Big Red Ant Found in Guatemala Which Lives on the Cotton Pest." The honorable Jas. Wilson, then secretary of agriculture, brought Cook's discovery to the attention of President Teddy Roosevelt. In the following months, the U.S. Congress made a special appropriation of $250,000 for continued investigations into cotton diseases and the study of weevil parasites, as well as the inspection of cotton products. The following year, $45,000, from a special fund of the secretary, was reserved "for work with Guatemalan ant and other possible emergencies."[34]

In July 1904, Cook arrived at the Department of Agriculture field station in Victoria, Texas, with about four thousand *Ectatomma* ants, in eighty-nine distinct colonies, that he had collected in Guatemala. The colonies were divided up for study at a host of laboratories around the country.[35] While Cook traveled to Washington to link his vision to the priorities of more established Department of Agriculture officials, a host of men set about studying the needs and interests of this insect. They described the slaying of the boll weevil with intimate attention to detail: "The ant's mandibles are large enough to grasp the weevil around the middle and pry apart the joint between the throat and the abdomen. The long, flexible body [of *Ectatomma*] is bent at the same time in a circle to insert the sting at the unprotected point, where the weevil's strong armor is open."[36] A preliminary outline for work with the ants included studies of adaptability to various soil conditions, the rate of egg deposition, the production of queens, and the conditions of mating. They set out to answer a series of questions: Can the production of queens be forced? Are stores of food gathered in special galleries? Will ants collect weevils resting quietly in squares?[37]

In trying to optimize the productive capacity of *Ectatomma* colonies, in working to increase the usefulness of this species to humans and its integration into economic systems, Cook set about studying every life stage of the organism (with barely acknowledged help from his wife).[38]

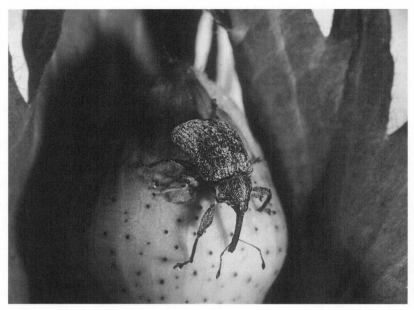

FIGURE 2.5. The boll weevil is a small beetle—about one-quarter inch long with wings and a very pronounced snout—that feeds on cotton buds and flowers. Indigenous to Central America, this insect began to spread into the United States at the turn of the twentieth century. The plantation economy of the southern United States was endangered as this tiny animal attacked cotton crops. Image source: USDA ARS Image Number K2742-6.

Mrs. Cook became captivated by the young larvae of the ants—small and plump grubs, like white sausages with distinct heads. Stout antennae and papillae frame the hard plates of their heads, allowing them to blindly taste and grope their way through the world. Seeds, dead insects, and animal matter, she noted, were collected by adult foragers and brought to the legless and seemingly helpless young. "With mouth parts adapted for eating out the soft interior tissues of insects," Mrs. Cook reports, the larvae of *Ectatomma* have "long, flexible necks [enabling] them to reach inside and clean out the sections of boll weevils laid by the workers carefully on the fat stomachs of their baby sisters."[39]

Mrs. Cook's thick descriptions of *Ectatomma* larvae afford an opportunity to consider the countervailing interests at play within ant colonies. Adult ants are only able to eat solid food in concert with their anatomically flexible youngsters. With ultrathin waists, called petioles, adults cannot move solid food into the digestive organs of their own abdomens. The larvae of ant colonies are thus agents of interessement—obligatory points of passage for food that stabilize networks of adults living together in the same nest or colony.[40] The embodied differences of adults and the larvae thus keep them interested in one another. With a conjoining of diverse body parts, with an intermingling of mutual utility and perhaps pleasure, adult workers and larval ants often eat solid food together. Chopping up the food with their mandibles, adults position manageable tidbits within reach of larvae. Ingesting bits of food, and excreting enzymes to predigest other solids, the larvae break the food down into chemical components. Larvae of many ant species secrete nutritious liquids that adults, in turn, drink.[41] Sometimes adult ants, in their excitement, will bite the larvae and drink the bodily fluids of their young.

Fluid exchanges of material and semiotic elements, liquid food containing chemical information capable of generating behavioral and bodily transformations, structures the social worlds of ants.[42] Human social worlds, according to a classic definition from sociology, involve collaborating and doing things together. They are communities of practice and discourse engaged in collective action. While much of the literature about humans is preoccupied with the roles of entrepreneurs, agents that are viewed as being central in the construction of social worlds, it is clear that a multitude is involved in the coproduction of ant worlds.[43]

EXCESS

Dr. O. F. Cook's imagination was captured by some of the same social behaviors that I observed nearly a century later. Cook found that neighboring *Ectatomma* colonies had low levels of hostility to one another. "The power of ants to distinguish at once between members of their own and of other colonies has long been recognized as one of the most remarkable refinements of instinct," wrote Cook in 1905. Whereas most ant species exhibit open violence toward individuals from other colonies, he found that ants from neighboring *Ectatomma* nests would often tolerate one another. "Members of two colonies will forage on the same cotton plant or tree trunk with no signs of animosity," he reported. "Stranger ants introduced into captive colonies for observation have not been attacked. They usually receive little attention; if they enter the burrow they are likely to be brought out and carried to the boundary of the enclosure, but are released without injury."[44]

Working at the Smithsonian laboratories in 2008, I became determined to better understand the behavior of these moving animals that were switching among social worlds. Joan Fujimura suggests that the science wars of the 1990s were "not about science versus antiscience, not about objectivity versus subjectivity, but about authority in science: What kind of science should be practiced, and who gets to define it?"[45] Keeping Fujimura's words in mind, I set out to study the diverging values and obligations structuring relationships among individual ants. Assembling an experimental apparatus out of found objects and specialized equipment—plastic tubs, petri dishes, dental cement, aquarium tubes, a slippery substance called fluon, and a Sony digital video camera—I worked to produce an experimental matter of fact, that members of distinct *Ectatomma* colonies exchange liquid food via trophallaxis.[46]

As an undergraduate, my research on *Ectatomma* had been restricted to watching solid food move among nests above ground. This movement of solid food, destined to be eaten by the larvae underground in different nests, was difficult to interpret. While some reports in the literature characterized it as thievery, I was convinced that there was more to this story.[47] Studying trophallaxis, fluid exchanges of material and semiotic elements, in a laboratory setting would enable me to trace actions oriented to the care of other beings, to consider how *Ectatomma* ants produce common worlds.

The term "trophallaxis"—deriving from the Greek words for nourishment and interchange—was coined by William Morton Wheeler in 1918. Wheeler, who was then the curator of invertebrate zoology at the American Museum of Natural History, carefully studied trophallaxis in diverse ant species and developed an elaborate model of the origin and continued functioning of insect societies.[48] He argued that "the origin of the behavior of individual ants *within the context of the colony* could not be explained in terms of individual inheritance. Mutual feeding relations were the true and necessary cause of social forms of life."[49] At least since the time of Wheeler's writings about trophallaxis, biologists have drawn analogies between the productive capacities of human societies and those of social insects—comparing the ability of human workers to earn wages to the ability of ant workers to collect food; comparing the collective wealth of a nation to the amount of energy stored in the bodies of workers or in caches of food; comparing systems for producing commodities to systems for reproducing new ant queens.[50] These comparisons have been grounded in economic models of rationality and scarcity.

Wheeler based his ideas of ant society on the writings of Vilfredo Pareto, an Italian economist of the early twentieth century, who in his early work assumed that human beings act rationally in pursuing their economic ends. Later in life, Pareto studied celebrations of great occasions, jubilees, graduation ceremonies, religious ecstasies, and excesses of all kinds.[51] Pareto suggested that human proclivities for these excesses were evidence of what he called "residues," forces that were distinct from instincts or biological drives. In William Morton Wheeler's hands, Pareto's work on residues was inflected with functional evolutionary explanations. Wheeler suggested, "The residues of the common man condemned him to a life that was functionally similar to the ant's."[52] Paging forward from Pareto to Georges Bataille—the librarian, surrealist, pornographer, and writer who argued that the exhaustion of excess is fundamental to the stability of societies—offers a window into worlds that generate more surprising possibilities.

Bataille's basic premise is quite straightforward: "The energies of the laborer are not completely exhausted (utilized) in the labor process itself. Surplus value represents the measurable portion of the worker's productive capacity which does not return to him or her as a wage. There is, however, another surplus, an unmeasurable excess, which does not return to the production process but is expended 'unproductively.'" Humans are compelled to destroy this surplus, according to Bataille, through large-scale festivals or wars. Bataille made similar arguments about other species: "The living organism ordinarily receives more energy than is necessary for maintaining its life; the excess energy (wealth) can be used for the growth of a system (e.g., an organism); if the system can no longer grow, or if the excess cannot be completely absorbed in its growth, it must necessarily be lost without profit; it must be spent, willingly or not, gloriously or catastrophically."[53]

As massive cargo vessels and cruise ships passed nearby at dizzying speeds in the Panama Canal, as I beheld the surpluses of the modern world system, I set out to design an experiment about the exchange of excess liquid food among *Ectatomma*. Collecting *Ectatomma* colonies for this experiment involved a visit to a festive space, a place where the value-added excess of late capitalism was routinely consumed. In a fragment of forested land next to a waterfall in El Giral, a small farming community about an hour outside of Panama City, I uncovered six *Ectatomma* nests among empty packaging of two brands of chocolate chip cookies (Choki's and Creamas Cuky), a supersized Cheetos bag, and some discarded wrappers of Papitas, a cheese-flavored snack. Amid this plastic litter left behind by human picnickers, and leaf litter from a huge *Pseudobombax* tree, I found a lively patch of *Ectatomma* ant nests. Between a discarded cardboard case of Miller Genuine Draft, as well as Balboa and Panama brand

FIGURE 2.6. Cargo ships in the Panama Canal. Photograph by Keith Dannemiller/CORBIS.

beer cans, I discovered a red bottle cap labeled, "A Product of the Coca Cola Company," with a cryptic message printed inside: SIGUE PARTICI-PANDO (keep participating).

After having a picnic of my own in El Giral with friends—Daniela Marini, an Argentinean ecologist who earned a master's degree from Yale's Forestry Program, and Jesus Hernandez-Montero, a bat special-ist from Mexico—I enlisted their help in observing and recording the transfer of food among *Ectatomma* nests. Amid human surplus, in this place where the excess fructose corn syrup and grain from North America and elsewhere was being expended in celebrating minor occasions and jubilees, we found that these ontological amphibians were flourishing. Crumbs left by picnickers, small insects, and small protein-packed snacks called Müllerian bodies, treats given to insects by *Cecropia* plants, were being carried from one *Ectatomma* nest to another. Worker ants were ex-iting the entrance of one colony and marching, usually unmolested, into the entrance of another colony.

After unearthing three colonies in El Giral, we transported them, in a U.S. government vehicle, back to the Smithsonian laboratories in con-verted Panama Canal Company buildings. After attaching two nests to a common foraging arena and giving the ants a week to adjust to their new circumstances, I let the paired colonies interact. Inside this experimental apparatus, I duly observed and recorded trophallaxis among the neighboring colonies I collected from El Giral—workers holding drops of sugar water opened their mandibles, retracted their mouthparts, and fed workers from another colony who gently antennated the donor's face, mandibles, and labium.[54] When I paired a colony I collected from El Giral with one from nearly ten miles away in the Canal Zone, I initially observed aggression among the ants—they were biting and dragging each other around the foraging arena. After growing accustomed to each other, after about a week, these unrelated ants started venturing into each other's colonies, and eventually feeding each other by trophallaxis.

These observations do not yet constitute a scientific fact—at this point there is a sample size of two paired colonies. If these observations can be replicated in other colonies, then it will be clear that *Ectatomma* workers are not just engaging in thievery, as some have suggested.[55] Perhaps members of distinct *Ectatomma ruidum* colonies engage in reciprocal altruism, giving gifts that demand future gifts in return. Finding that individual ants seem to be rational economic actors, like a long list of other animals—lions, crows, and baboons, for example—would confirm deeply held assumptions of many biologists. Perhaps, though, these creatures do not have good economic sense. Following a large body of literature reporting economically irrational behavior in humans, some breaches of rationality in animals have also been described.[56] Further research with *Ectatomma* might reveal that their gifts of liquid food happen according to fleeting whims, sentiments about the distribution of surplus that escape rational calculus.[57]

SIGNIFICANT OTHERS

If ants enfold each other into common worlds with fluid exchanges of material and semiotic elements, if adults and larvae are agents of interessement, then perhaps they also care for other species of beings and things. The lives of *Ectatomma* ants are entangled with plants that secrete sugary liquid offerings, phloem-sucking leafhoppers that exude honeydew treats out of their anus, and caterpillars that communicate with the ants in high-pitched stridulatory sounds.[58] To play with Heidegger's language, *Ectatomma* workers are captivated (*benommen*) by other beings and are becoming with others in common worlds.[59] These enterprising ants are constantly forming and transforming emergent ecologies and assemblages. While wandering within the cosmos, individual *Ectatomma* ants form political articulations with individual plants and animals. Building cosmopolitical worlds—together, tooth and nail, with other organisms—ants form stable, but contingent, relations against the backdrop of the unknowable beyond.

Dr. O. F. Cook's work offers ample evidence that *Ectatomma* ants are not trapped within ontological cages. In one *Science* article, he wrote, "The insect is not, like some of the members of its class, confined to a single plant."[60] "The species of *Ectatomma* are widely distributed, enterprising ants," he wrote in a separate article. "Instead of being a rare 'archaic' curiosity, it is decidedly the . . . most abundant insect of the Guatemalan cotton fields."[61] But Cook also learned a painful lesson—that the worlds of *Ectatomma* are not endlessly expansive. He "planted" dozens of *Ectatomma* colonies in the cotton fields of Texas during the summer of 1904. Most of these colonies died over the winter. By March 1905, Cook was not expecting any of the few remaining ants to survive beyond the month.[62] As he failed to serve the interests of the ants under his care, the media raged with allegations of financial mismanagement, and his supervisors were forced to respond to "sore questions."[63]

Since Cook's time, other investigators have found this ant tending the extrafloral nectaries of many other plant species besides cotton: for example, on woody liana vines (*Dioclea elliptica*) in the canopy of a lowland Amazonian rain forest of the upper Orinoco and on saplings of a tree in the legume family (*Stryphnodendron microstachyum*) on the Caribbean slope of Costa Rica.[64] On *Ingas*, a kind of tree that is popular in reforestation initiatives since they draw diverse forms of life together in a living architecture, *Ectatomma* ants are often seen at nectaries alongside beetles, butterflies, wasps, and flies (see chapter 10, "Possible Futures").[65] At the same time as *Ectatomma* serve as convivial members of multispecies communities, they can also be pesky parasites. Using a particularly clever trick, some *Ectatomma* sniff out the pheromones of other ants, smaller species like *Pheidole*, and follow their chemical trails to sources of food.[66]

Diverging values and obligations structure ambivalent relationships between ants and plants—cosmopolitical articulations characterized by mutual utility and mutual exploitation. Douglas Altshuler has found that the presence of my favorite ant has certain positive effects for *Psychotria limonensis*, a common shrub in the forest understory of Central America. *Ectatomma* foragers increase the rate of pollination for this species—likely because they startle pollinators, like butterflies, making them move to other plants. Ants also serve the interests of *Psychotria* by defending the plant from herbivorous insects and preventing the loss of ripening fruits. The cosmopolitical world of *Psychotria* also includes

FIGURE 2.7. Treehopper nymphs feed on sap from plants by piercing the stems with their beaks. Excess sap, concentrated in honeydew, is exuded out of the nymphs' anus, and this sugary liquid often attracts ants. In this picture, *Ectatomma tuberculatum* is tending a treehopper nymph in the Jatun Sacha Reserve in Ecuador. Photograph courtesy of Alex Wild.

fruit-eating birds—tanagers, manakins, and neotropical migrants—that eat ripe fruits and disperse the plant's seeds. Even if both *Psychotria* and *Ectatomma* have cause to be interested in each other's continued existence, the ants do not always act in the best interest of the plant and its avian companions. Ants scare off fruit-eating birds. After fruits ripen, the continued presence of ants thus does not serve the assumed interest of the plants in seeding new territory.[67]

While jealously guarding their plants from flighty interlopers, *Ectatomma* ants remain open to overtures from other entrepreneurial agents—creatures that work to enlist them in competing cosmopolitical worlds. "Adding insult to herbivory," in the words of Philip J. Devries, *Ectatomma* ants sometimes welcome leaf-eating caterpillars to feast alongside them on plants with extrafloral nectaries. These caterpillars have noise-making organs that attract *Ectatomma* and other sorts of ants. The sounds made by the caterpillars average at 1,877 hertz, which would be audible to human ears if they were not so very faint. Their repertoire ranges from simple sounds, like "bub . . . bub," to fancier noises such as "beep ah ah ah beep" and "biddup . . . biddup . . . biddup." Caterpillar calls summon ants to their defense against predatory wasps and parasitic flies. As a reward for responding to the summons, the caterpillars secrete a liquid gift—a nutritious liquid that is significantly higher in amino acid concentrations than the plant nectar. *Ectatomma* ants tend the caterpillars "with greater frequency and fidelity" when compared to the plant.[68]

Lori Gruen's notion of entangled empathy might help explain why ants have greater fidelity for caterpillars than for plants.[69] Entangled empathy is not a mere instinctual response, but involves a commitment to the well-being of others—an awareness of others' interests and a motivation to satisfy those interests. Gruen uses empathy to understand multispecies entanglements—specifically her own interactions with chimpanzees. Exporting these ideas beyond our own situated perspectives, the embodied umwelt of primate vision, contains the danger of imposing anthropomorphic assumptions on other worlds. Even still, Gruen's work prompts me to ask: Do ants perceive the interests of the plants they protect? Do they recognize plants as beings in the world? Quite possibly not. Are ants aware of the caterpillars' interests and are they motivated to fulfill them? Quite possibly yes. With intriguing sounds and an anatomical structure similar to ant larvae, it seems plausible that these caterpillars appear to *Ectatomma* as beings (cute baby insects) that demand empathetic regard rather than as things to be jealously guarded.

Departing from the notion of superorganism, which has been applied to both ant colonies and ecosystems, I suggest that we understand such multispecies communities as ensembles of selves—associations composed of conscious agents who are entangled with each other through relations of reciprocity and accountability, who regard each other with empathy and desire. The notion of ensemble is borrowed from Paul Kockelman, who in turn has purloined William James's ideas about the self—the sum total of things we call our own. Selfhood involves what constitutes part of the ensemble. In human realms, the self-as-ensemble includes one's clothes and house, one's ancestors and friends, one's nail clippings and excretions, one's body, soul, thoughts, and ways of being in the world. Actions oriented to the care of beings and things enlist them in the ensemble.[70] "To care for others is to care for one's self," writes Deborah Bird Rose in a related vein. "There is no way to disentangle self and other, and therefore there is no self-interest that concerns only the self."[71]

"Selfhood is not limited just to animals with brains," adds Eduardo Kohn. "Plants are also selves."[72] Even if ants do not recognize plants as beings in the world, they are perhaps being unwittingly enfolded into the self-defense system of the plant.[73] When an *Ectatomma* ant resting on a plant sees a large vertebrate, a potential predator or herbivore like me, she will often turn her whole body to face off—jaws open, legs firmly planted, stinger ready. If these persistent threats are empty (any *Ectatomma* aficionado knows that these ants have difficulty stinging humans and will scurry away, and try to hide, upon serious molestation), they serve as poignant warnings against an uncritical celebration of multispecies mingling. These animals are not only "good to think" (as Lévi-Strauss suggested), or good "to live with in the flesh" (as Donna Haraway added), but can also sting and bite.[74]

Persistent reminders of the unease generated by my fondness for their kind prompted me to take pleasure in the confusion of some boundaries, while taking responsibility for constructing others.[75] While learning how to see ensembles of living selves in worlds around me, I found that caring for some beings or things is best done from a polite distance. Amid fleeting moments of reciprocal capture, involving ants who were constantly trying to escape, I found gaps in our gaze, disjunctures in our interests.

After releasing all the ants in my care from the lab, I ventured beyond the realm of the Smithsonian Tropical Research Institute, a social world of ecological scientists where my love for *Ectatomma* was unremarkable. In the Reverted Zone, the restricted lands around the canal that reverted to Panamanian sovereignty in 1999, I found *Ectatomma* ants foraging in the shadows of abandoned satellite dishes, collecting dead insects under electric lights, and living in an expansive network of nests in neatly manicured lawns. Few of my fellow humans were articulate about the ants living in the grass all around them. While more than one of my interlocutors looked at me as if I were a little off, for initiating a conversation about insects, I found that strangers were more than happy to talk with me about their hopes, desires, and dreams. In the Reverted Zone, I began to study the contours of biocultural hope in a blasted landscape. Certain forms of life were proliferating in the aftermath of military and commercial disasters, anchoring dreams about a posthuman utopia. Hope for other organisms, endangered species, was emerging as they were being enfolded into artificial ecosystems built to protect them from agents in the environment.

CHAPTER THREE

HOPE IN THE REVERTED ZONE

. . .

Living as an ethnographer in the City of Knowledge—formerly Fort Clayton, the one-time command-control-intelligence center of the U.S. military's Southern Command—I began studying emergent forms of life in the abandoned architecture of empire. My temporary residence in September 2008 was an army barracks converted into a hostel for road-weary backpackers on the gringo trail. Formerly off-limits to all without proper authorization, Fort Clayton had become a suburban enclave populated by middle-class Panamanians, indigenous Guna from the Caribbean coast, the staff of international organizations, and a few remaining Zonians, U.S. expatriates who were born in the Canal Zone. On my first day there, I took a late-afternoon bicycle ride in misty rain, alongside many other residents engaged in the pursuit of physical fitness. An aerobics instructor was screaming out chants at the top of his lungs to a group of women doing exercises on big inflatable balls inside a huge Kiwanis Club gymnasium. A pair of men, pitcher and batter, were at work in a nearby cage. Joggers and many other bikers hailed me with smiles, nods, and lifted eyebrows—recognizing me as a fellow recreator and a possible neighbor.

During this bike ride I encountered prolific forms of life that were flourishing amid shifting political and economic forces. Agoutis, large herbivorous rodents, were running through the suburban lawns as I whizzed by. Stopping my bike by the side of the road, I found lively *Ectatomma* colonies in the closely cropped grass. Chain-link fencing separated this yard from a large culvert where a caiman was sitting on a log. As dusk settled in, a chorus of frogs started calling from the culvert. The mating call of one species, the túngara frog (*Engystomops pustulosus*), could be easily distinguished from the rest. Jon Fuller, a frog biologist who first visited Panama in 1980, during the halcyon days of the U.S. military occupation, described the calls of túngara frogs in colorful terms: "They sound like the gun battles in Star Wars between the Storm Troopers and the Good Guys!"[1] Stanley Rand, a Smithsonian staff scientist who dedicated much of his professional life to understanding the complexity of these beautiful calls, described their two distinct elements in more precise terms: "One sounds like a descending tenor groan or whine. The other is an explosive call, shorter and higher pitched than the first, giving the subjective impression of a clack, or, when rapidly repeated, a rattle."[2]

Following túngara frog calls around the City of Knowledge on my bicycle, I found large groups of frogs in ephemeral rainwater ponds sitting in parking lots. They were also thriving in gutters and ditches built decades earlier by the Army Corps of Engineers. Some frogs were strategically positioning themselves in drain pipes and other acoustically enhanced infrastructure that noticeably amplified their calls. Bringing my casual observations together with information I had gleaned from scientific archives, I surmised that túngara frogs had been constantly moving among multiple elements—living on earth and in water, choosing among environmental worlds and human architectures. In short, these frogs had become ontological amphibians. Alongside *Ectatomma* ants and agoutis, túngara frogs were flourishing alongside people, within architectures designed with the well-being of others in mind. Switching from one element to another, these native animals had become adept at invading and occupying emergent worlds.

While living in the City of Knowledge, I heard about multispecies communities that were flourishing in zones of abandonment, in blasted landscapes that had been deserted by the U.S. military.[3] Bombing ranges pockmarked with blast craters—riddled with unexploded ordnance as well as chemical agents like VX gas (a nerve agent)—were transferred to Panama on December 31, 1999, when the United States ceded its territorial claim over the Canal Zone. The U.S. Department of Defense declined to pay for the cleanup of these areas, which contained more than fifty thousand unexploded bombs spread across some 1,665 acres.[4] Despite a long history of use as a bombing range, one parcel was characterized by the military as "relatively pristine, mature, tall-statured evergreen forest with ecosystems that have not been significantly disturbed by human activity."[5] The Department of Defense commissioned a study by the Nature Conservancy that concluded, "DoD lands harbor rare plants, animals, and forest types that are quickly vanishing in Panama, and are considered globally rare by the scientific community."[6] The Nature Conservancy argued that cleanup activity could result in deforestation, habitat loss for wildlife, and long-term environmental impacts.

FIGURE 3.1. Unexploded ordinance often remain undetected for decades. Photograph by AusAid.

When people rally to save "the environment," perhaps one should always follow Donna Haraway and ask, "What counts as nature, for whom, and at what costs?"[7] Here, in these forsaken parts of the Reverted Zone, a state of nature was being protected that directly served U.S. political, economic, and military interests. Preserving this vision of nature came with a direct cost to Panamanian lives and subsistence practices. The former U.S. bombing ranges were cordoned off, marked with signs in Spanish noting that more than twenty people had already been killed after venturing beyond. For some this risky situation presented an economic opportunity. A Panamanian consulting company, Isthmian Explosive Disposal, began working on the margins of former U.S. bombing ranges. They cleared broad swaths of forest of unexploded ordnance while making way for commercial enterprises like development at Tocumen International Airport and the expansion of the Panama Canal.[8] Areas with the highest impacts from U.S. target practice, places with tens of thousands of unexploded bombs, will probably remain outside of commercial expansion indefinitely. In this "undisturbed" forest, endangered animals—like spider monkeys (*Ateles geoffroyi*), Baird's tapirs (*Tapirus bairdii*), and cotton-top tamarins (*Saguinus oedipus*)—are free to proliferate and run wild.[9]

Instead of returning to the pristine state of Nature, as imagined by the military, these forests of Panama's Reverted Zone have entirely novel features when compared with forests that came before. New rivets— unexploded bombs and chemical weapons—have fastened an emergent ecosystem in place.[10] These rivets have the indeterminate properties of the *pharmakon*. Classically, the pharmakon is a poisonous substance that can have a therapeutic effect, depending on the dose, the circumstances, or the context. The poison of the chemical weapons left in this forest offered a possible cure for ecological communities that were struggling to survive the ills of commercial development. Isabelle Stengers understands the pharmakon as any obstacle that presents an opportunity—something with unstable attributes, with effects that can shift from one extreme to another. Unexploded ordnance in Panama certainly has "pharmako- logical" properties. Bombs usually act as hidden talismans that protect wildlife from people. But sometimes the attributes of these unstable ob- jects suddenly shift, with blasts generating violent pain and death for the animals and plants that have found a home in these abandoned lands.[11]

Some environmentalists, strangely, regard such spaces as utopias. Works of popular fiction describe a future where other species will flour- ish without us in the aftermath of disasters. Alan Weisman's *The World without Us* (2008) describes how animals, insects, plants, and mold will inhabit our homes, making them unrecognizable. *Life after People*, the television documentary by David de Vries (2008), features a biologi- cal scientist who imagines "lovely" postapocalyptic scenes of New York City covered with vines, with hawks sailing around, with "fire hydrants sprouting amidst cacti." This documentary also visits the "ghostly vil- lages" surrounding Chernobyl, which was abandoned by humans after the 1986 nuclear disaster, and is now home to packs of radioactive wolves and moose. Creatively blending fiction and nonfiction, these authors and filmmakers avoid the catastrophic mayhem of a human holocaust by de- scribing "the abrupt exit of humanity from the scene, leaving all other species and the environment intact."[12] A self-loathing perspective rather than a self-critical perspective, according to Vincent Geoghegan, drives dreams about these posthuman utopias.[13]

While living in the City of Knowledge, I found competing utopic dreams. If the hopes of some environmentalists were pinned to the de- parture of all humans from abandoned bombing ranges, the collective dreams of Panamanians were tied to the relative absence of a more spe- cific figure: the gringo.[14] During the long decades of U.S. military occupa- tion, the eighteenth-century Spanish noun *gringo* was playfully developed into a command, a protest slogan directed at troops in jungle fatigues: "Green Go Home!" U.S. soldiers stationed in Panama became "sticky objects" gathering up the bad feelings about diverse problems—economic inequality, human rights abuses, failing water infrastructure, access to land for farming and hunting.[15] The occupation seemed like it would never end, even after President Jimmy Carter and Panamanian leader Omar Torrijos signed the Panama Canal Treaty in 1977, promising the handover.

Panamanian dreams from this earlier historical period might be best understood with help from Jacques Derrida, who makes a useful distinction between apocalyptic and messianic thinking.[16] Messianic hopes proliferate against the backdrop of seemingly impossible odds, amid dashed dreams and disappointing circumstances.[17] Looking to possible futures, rather than to absolute endings, Derrida suggests that these hopes contain "the attraction, invincible élan or affirmation of an unpredictable future-to-come."[18] "Not only must one not renounce the emancipatory desire," insists Derrida, "it is necessary to insist on it more than ever."[19] During the long decades of the twentieth century, a homogenous, empty time in Panama when power functioned predictably, people held on to dreams of a better tomorrow. Rather than the apocalyptic dreams of environmental writers who imagine a future without us, Panamanians pinned their hopes to an imagined event: the moment when the gringos would march through the gates of the airports and harbors, never again to return. As December 31, 1999, approached on the horizon—the promised handover event that coincided with the coming of a new millennium—a sense of homogenous and empty time quickly gave way to a sense of messianic time.[20]

As the troops pulled out, an emergent middle class in Panama was able to enjoy newfound freedoms. Dreams of prosperity and security—seemingly impossible in an earlier era—came within reach. As Panama became an emerging market, as a new class of consumers enjoyed a stable and comfortable lifestyle, the planned communities created for the white U.S. government workers with the "gold" classification were taken over by diverse communities. Afro-Caribbean intellectuals, indigenous artisans, and mestizo entrepreneurs all moved to the Reverted Zone. A modern utopia arrived. The magical realism animating earlier dreams in Latin America—the Macondo described in Gabriel García Márquez's novel *One Hundred Years of Solitude*—was replaced by McOndo, the enchanted domain of McDonald's and modern wonder.[21]

In September 2008 I enjoyed the fruits of this utopia during my bike

FIGURE 3.2. The Panama Canal Zone became the Reverted Zone in the year 2000. Killbyte, a Zonian who is now living in Florida, took this picture of a former Canal Zone clubhouse in 2009 when she visited Panama for her thirty-fifth high school reunion. This establishment was once run by the U.S. government and was the only store, restaurant, and movie theater in the old town of Diablo. In the words of Killbyte, the clubhouse has become "some horrible, foreign, bizarre nightmare in reality and my dreams!" (https://www.flickr.com /photos/killbyte/4613230214). Photograph by Teresa DuVall.

ride through the City of Knowledge. Parking my bike outside a Subway restaurant that was in the process of locking up for the night, I grabbed dinner while participating in (and observing) an emerging market of consumer capitalism. They had a bilingual menu and monolingual, Spanish-speaking staff. I wavered between a new sub that was advertised on the wall, which featured kosher provolone, and the traditional Veggie Delight. In the end I went for the veggie sub on honey oat bread, which I ate while scrolling through the latest news stories about the global financial crisis on my smartphone. At a moment when reports of scarcity were looming large on the international stage, I wondered about the present state of affairs and the future: was I sitting in a new space of plenty that free-market economics had long predicted, or on the brink of a tragic epoch, in which anarchic market forces would generate new waves of socio-ecological changes?[22]

While living in the City of Knowledge, I became friendly with Jayne Reyes, a young Panamanian woman about my own age who owned and operated the backpacker hostel where I was staying. She told me about how she was able to actualize her own modest dreams after the departure of the American soldiers. She also told me about the horrors of military occupation—about lived experiences of terror in the days leading up to Christmas 1989, when George Bush senior directed U.S. troops to bomb Panama City and depose President Manuel Noriega. After the soldiers left, Jayne began enjoying a comfortable lifestyle within their former barracks. Renting out rooms to travelers enabled Jayne to pay her bills and still have enough money to occasionally eat out. If some of her contemporaries had pinned their hopes on the disappearance of all gringos, Jayne was finding opportunities in our continued presence.

Perhaps hope itself has the properties of the pharmakon.[23] Obstacles to freedom, like gringos who keep coming back, can be reimagined as opportunities. The U.S. Embassy, with a host of Department of Defense personnel, became one of the largest employers in the Reverted Zone, offering Panamanians jobs. International organizations—like the Red Cross, the United Nations, the Organization of American States, and the Nature Conservancy—also set up offices within the architecture of American imperialism, in the abandoned military buildings of the City of Knowledge.[24] Competing utopian projects—demilitarized spaces of hope that were devoid of humans and suburban communities where people were living middle-class dreams—existed uneasily, side by side.

In the fall of 2008, I found some Latin Americans were again captured by "hope"—a keyword of the Obama presidential campaign. As Obama worked to embody the messianic spirit—quoting Martin Luther King Jr.'s famous plea about "the fierce urgency of now"—people all over the globe imagined themselves inhabiting the slippery pronoun "we" in "Yes We Can!"[25] Optimism can be cruel, according to Lauren Berlant, when you discover that the dreams you are attached to are either "*im*possible, sheer fantasy, or *too* possible, and toxic."[26] Shortly after Obama was elected, the toxic properties of hope became apparent to many people who had once been caught up in the collective enthusiasm behind "Yes We Can!" The unstable word "hope" suddenly shifted from being an imagined cure to the ills of the United States to being a vacuous political slogan—a cruel and cynical slap to those who audaciously dared to participate in collective dreams.[27] While moments of messianic arrival are often disappointing, it is worth thinking critically about what one might do in postrevolutionary moments, after an event like the end of the military occupation in Panama or the election of the first African American president of the United States. Does one give up on hope after a particular moment fails to actualize every element of expansive dreams? Or does one appreciate what has shifted and maintain openness to the possibility of future changes?

When I returned to Panama in 2014, Jayne was waiting for me at the airport outside customs—looking up at a screen displaying flight arrivals. I put my baggage in her trunk, on top of a damaged hubcap that was sitting on the floor. Paying for parking, and then a series of tolls, Jayne sped down the road—maneuvering through shortcuts, around the back of a shiny new shopping mall, and out onto an overpass—cruising over tidal flats lined with mangroves. Skyscrapers poked up on the horizon—some freshly completed and others still under construction. "We still haven't felt the economic crisis," Jayne said in Spanish. "These new buildings are going up faster than trees. Maybe it is related to the expansion of the canal." As we wound around the skyscrapers, past manicured oceanside parks that looked like they belonged in Monaco or Los Angeles, she recounted Panama's recent presidential elections. A supermarket chain owner, a multimillionaire, won. As the overpass dumped us out on city streets, and into traffic, Jayne reached across me to lock my door. "The *locos* could try to open it," she said. "So there is still poverty?" I asked. "Yeah, maybe there always will be," Jayne answered.

A few days later, I tried to visit the Subway where I had purchased my Veggie Delight some six years before. But I found that the chain restaurant was gone. The parking lot had been broken up. An austere black reflecting pool and a looming wooden structure—the first building in Panama given the LEED gold certification for environmental design—stood in its place. New businesses included Jungle Juice, Itsaski Gourmet Marisquería and Cevichería, the Health Store, and T-Bar—tea and fusion food. A sushi restaurant served me edamame and sashimi in a diverse assortment of plastic containers along with Dasani water, bottled by the Coca-Cola Company. On the site of the middle-class utopia I had encountered in 2008, I found that the crisis tendencies of capitalism had widened and deepened economic gaps in Panama. Ecological, spatial, and cultural forms had once again been reconfigured into a new world that embodied cosmopolitan values.[28]

If Panamanians were united in their dreaming in the 1990s when I first worked at the Smithsonian labs, pinning their hopes on the departure of U.S. troops, most people I met in 2014 were less articulate about their desires. Rather than envisioning sweeping social and political changes, I found people talking about their desires for professional advancement, about wanting to buy a new house or a car, about dreams of entering heaven or winning the lottery.[29] Religious dream worlds and the desires of consumers became a backdrop as I studied thinkers and tinkerers who were doing imaginative and practical labor to keep noteworthy forms of hope alive. Amid ongoing disappointments and disasters, at the intersection of warring utopic visions, these organic intellectuals were grounding hopes in plans for a shared future with endangered creatures they loved. I began hearing about people whose hopes were tied to the well-being of frogs.

While multiple species of ontological amphibians—like the túngara frog and *Ectatomma* ants—were proliferating in the Reverted Zone, many other species of literal amphibians were becoming imperiled as a result of commercial development, agricultural pesticides, and emergent diseases. Drastic declines in frog populations were noticed worldwide starting in the 1980s—particularly in Australia and the Americas. Upward of 165 species of amphibians are believed to have gone extinct in recent years. About 1,895 amphibians, over one-third of all described frogs, salamanders, and caecilians, became threatened with extinction.[30] Many different hypotheses emerged to explain these sudden declines—global warming, deforestation, habitat loss, pesticides—all putting humans at the center of the story. Anthropocentric stories gave way to a more complex multispecies story when a new microscopic pathogen, a chytrid fungus, was described in 1999. Conservationists began building a global assemblage of biosecure holding facilities and cryogenic banks, the Amphibian Ark, to protect endangered animals from the fungus. Mechanized ecosystems, architectures of immunology, were assembled to hold endangered animals in perpetuity. As fungal parasites found clever new ways to invade the body of amphibian hosts, people began folding frogs into novel technological and symbiotic assemblages. Visionaries grounded hopes by caring for living figures: multiple species of endangered frogs.

Figures are graphic representations, like drawings. "To figure" also means to play a role in a story, notes Donna Haraway.[31] Figures gather up collective desires into stable objects of hope, serve as anchoring points for dreams, and produce the conditions for collective action.[32] Messianic figures, like Jesus Christ or President Obama, embody expectations of sweeping changes—playing a role in visions of ultimate salvation and apocalyptic destruction, offering promises of radical transformation, only to disappoint even if they do arrive. Talking with caretakers who worked in the Amphibian Ark, I found people who were grounding more modest hopes in the living figures of actual animals, who were "disposed to take pleasure in the perfection, well-being or happiness of the objects of their

FIGURES 3.3 AND 3.4.
Biosecure shipping
containers and everyday
technologies—PVC
pipes, electrical outlets,
fluorescent lights,
modular shelving, and
aquariums—have been
assembled into life
support systems for
frogs. Photographs by
Eben Kirksey.

love," to borrow the words of Gottfried Leibniz.[33] Caretakers were working to guarantee the endurance, survival, and flourishing of endangered amphibians, protecting above all "the desire that made [these frogs] powerful enough to have magnetized an attachment to them."[34]

Just a few bus stops away from the City of Knowledge, within a facility run by the Smithsonian, I found a diversity of frogs living within retrofitted shipping containers, or "amphibian pods." Harlequin frogs of the genus *Atelopus*—some of the most brightly colored amphibians in the world—were the most celebrated residents of the amphibian pods in the Reverted Zone of Panama. Thirty species of *Atelopus* were already presumed extinct. Over fifty of the remaining species in this genus were experiencing rapid population declines.[35] Representatives of these endangered frogs were flourishing within new emergent ecosystems inside the shipping containers. Romantic trysts between a couple of the frogs in the days leading up to my visit had generated much excitement among the staff. Members of two endangered species, *Atelopus glyphus* and *Atelopus certus*, had produced hundreds of eggs, strung together like pearls on a necklace. "We were warned that we might not be able to keep these frogs alive," said Brian Gratwicke, the director of the Panama Amphibian Rescue and Conservation Project. "But through a little bit of guesswork, attention to detail and collaboration with other husbandry experts—we've managed to breed them."[36]

Some species are doomed to die because they are unloved by humans.[37] Interspecies love is not without its own problems. When "the category of 'endangered species' takes hold of organisms," writes Donna Haraway, it "subjects them to the ambiguous grace of salvation, specifically being saved through a regulatory and technological apparatus of ecological and reproductive management."[38] No cure for the chytrid fungus is in plain sight, and no ecological management plans have been able to stop its spread. With no promise of salvation on the immediate horizon, human caretakers of endangered frogs are learning to be effective *bricoleurs*—making creative and resourceful use of whatever materials are at hand. Caring for these animals involves learning along the way.[39]

Initially, caretakers with the Panama Amphibian Rescue and Conservation Project collected insects from surrounding grasslands and woodlands to feed their frogs. But in 2010 they suddenly lost many of their *Atelopus limosus*, dark harlequin frogs with green stripes from the Chagres River basin. After consulting with a specialist veterinarian and poring through their care sheet logs, they realized that the main difference between the treatment of the recently deceased animals and the other frogs in their care was their diet. They had been feeding wild-caught termites to their *Atelopus limosus*. After these frogs died, they stopped catching wild insects and switched almost entirely to raising their own food for the frogs. "In the United States you can just pick up the phone and order more fruit flies or mealworms with a credit card," said Angie Estrada, one of the keepers. "Here we have to raise everything ourselves."

Volunteering my time in the "fly room," a squat air-conditioned building with a single fluorescent light, I learned about the unusual forms of life that are proliferating in the Amphibian Ark's transnational network. I helped care for mutants that had been imported from the United States: a wingless strain of fruit flies (*Drosophila melanogaster*) and a flightless strain of *Drosophila hydei*, a slightly larger species. Robert Kohler, a devoted historian of fruit flies (*Drosophila*), has described "a proliferation of mutations in *Drosophila* [that] altered the domestic ecology of experimental organisms and disciplines." Many of these mutants—like flies with olive body color, or speck wing axils—were produced by intense inbreeding for obscure scientific experiments.[40] If many of these mutants are now confined to the annals of scientific history, others have proliferated as new uses for them have been found. The fly room was home to a multitude of tiny animals. They were being reared in clear plastic containers, filled one-quarter full with a moist fruit fly medium that had also been imported from the United States. The containers were sealed on the top with coffee filters so that the flies could breathe. My job was to hunt out containers with filters that had become wet, from the ambient humidity of the culture, and replace them. Filters coated with wriggling fruit fly larvae (aka maggots), had to be saved—pushed back into the containers—to increase the production of coming generations of flies.

From the fly room, the adults were transferred to a skinny zookeeper aptly named Lanky, whose daily routine involved feeding them to the endangered frogs. Lanky is a member of the Wounaan—an indigenous group from Darien Province, where many of the frogs were collected. After growing up in Panama City, where his father was a graphic designer, Lanky moved to San Antonio, the natal village of his wife, which is located in the Reverted Zone. San Antonio was founded in the 1960s when a young Wounaan man was invited to start training U.S. government agents in jungle survival skills. As I expressed interest, Lanky invited me to San Antonio for a visit. A few days after my first day in the fly room, I took him up on the invitation.

Lanky told me to wait in the Butterfly Garden of Gamboa Rainforest Resort, a luxury hotel overlooking the Chagres River, where I could get a boat ride to San Antonio. The Wounaan had a standing arrangement with the resort management that let them use a rickety dock with rotten wooden boards on the margins of the property. Guests who stay at the resort can participate in a variety of activities—fishing on the Panama Canal ($200), gliding through the rain forest on an aerial tram ($40), or visiting an authentic Indian village ($25). (The resort runs high-end fishing trips and tours of the Indian village from a nearby marina.) After I waited for a few minutes on the rickety dock, an aluminum canoe paddled by two teenaged Wounaan boys came to pick me up. It had the look of an old summer camp boat, perhaps left over from the U.S. occupation, with a faded number five gracing the bow. Switching out with one of the teenagers, I took a roughly hewn wooden paddle and sat up front.

We made our way through winding river channels that were choked with vegetation—water hyacinth, lily pads, and water lettuce—that made paddling difficult. The vegetation was also mixed together with trash—plastic bottles, coffee cups, shopping bags, engine oil containers—that had washed downstream, the teenagers said, from the Panamanian town of Chilibre. San Antonio was buzzing with activity as we paddled up—people were laying out clothes to dry on lines and swinging from hammocks strung in house rafters. Palm-thatched houses up on stilts had electrical extension cords snaking up to bare lightbulbs. A Peace Corps volunteer had recently installed two computers with a satellite uplink to the Internet in a small shed, next to a concrete outhouse. Lanky was waiting for me in a long house with a variety of arts and crafts laid out for display: necklaces made with the seeds of a liana commonly called *ojo de buey* in Spanish (eye of the ox in English, *Mucuna pruriens* in Latin), baskets woven out of reeds, and animals that had been carved out of tagua nuts (*Phytelephas* sp.).

As we started chatting about San Antonio, about the village's historical entanglements with the U.S. military, Lanky called over Felipe Cabesón—the son of the man who founded the village. Felipe obligingly agreed to be interviewed on the spot. Waxing eloquent about the bygone days of U.S. empire, he told me about some of his fond memories of "the students," the U.S. government employees who learned jungle survival skills from his father. Between 1962 and 1975, when the Tropic Survival School closed, more than eleven thousand government personnel trained here: from the U.S. Air Force as well as ambassadors, embassy staff, civilian scientists, and special agents. Astronaut John Glenn and moon walker Charles "Pete" Conrad are among the alumni of this program.

Identifying poisonous and nonpoisonous varieties of snakes and frogs

FIGURES 3.5 AND 3.6. Astronauts from NASA trained with Embera and Wounaan teachers during a Tropic Survival School program in 1963. A faded and time-worn photograph of these esteemed alumni graces a wall in the lobby of the Gamboa Resort Hotel. Neil Armstrong, John H. Glenn Jr., L. Gordon Cooper, and Pete Conrad are pictured with an unidentified trainer, who was preparing them to survive if a misguided reentry landed them "in a remote wilderness area," according to the NASA archives. Photographs by Eben Kirksey and the Johnson Space Center (Image # s63_08394).

FIGURE 3.7. Lanky Cheucarama pictured with some of his carvings and his raw material, tagua nut. Photograph by Eben Kirksey.

was part of the curriculum, according to Felipe. The students also learned about ways of moving through the forest undetected. One of his earliest memories is of sitting around the cook fire one day while his mother prepared a rich stew with taro, plantains, and iguana meat. He could not understand the language of the students, English, but from the expressions on their faces and the animated discussion that ensued, it was clear that some loved the stew, while others found it unpalatable. The hardcore students ate all sorts of things that the Wounaan consider inedible. Some even ate frogs—not the species that Lanky later worked to protect, but the common smoky jungle frog (*Leptodactylus pentadactylus*).

Over the previous ten years, life had become difficult for the Wounaan, as the North Americans left and the Panamanians took over, according to Felipe. While Panamanians had long imagined sweeping social and political changes with the departure of U.S. troops, the Wounaan had enjoyed special freedoms during the occupation.[41] Tangles of aquatic vegetation in the Chagres River now make paddling tricky. Trash, washed downstream from Panamanian settlements, was not there before. Previously, the Wounaan were allowed to live in the Canal Zone, while ordinary Panamanians were prohibited to enter. Now they were routinely being harassed by Panama's ecological police. "We can no longer cut down trees to make our clothes. We can't hunt in the forest," Felipe said. He spoke of transitions that were pushing the Wounaan away from relations with ecological communities, bringing them into ambivalent relationships with market economies. "Now we have to travel to Panama City to buy food at the supermarket. The resort used to bring tourists who each paid $25 to visit us, but then they started playing money games. Now we refuse to let them come here, so the tourists only visit the Embera Indians farther down the river."

After concluding the interview with Felipe, I browsed the table of handicrafts that Lanky had prepared for me. I purchased a collection of tagua nut carvings of frogs—a red poison dart frog (*Dendrobates pumilio*) and the red-eyed tree frog (*Agalinis calidryas*), as well as one Panamanian golden frog (*Atelopus zeteki*). Lanky was caring for other species of *Atelopus* in the biosecure pods of the Reverted Zone, creatures that are less a part of our cultural lives than the frogs depicted in his tiny sculptures. Holding on to this carving of *Agalychnis callidryas*, the Panamanian golden frog, I began following frogs from the Reverted Zone to the highlands of Panama and beyond. I started to study the ecologies emerging around the Amphibian Ark, the transnational network of zoos, aquaria, universities, and businesses united by the mission of saving endangered frogs.

CHAPTER FOUR

HAPPINESS AND GLASS

. . .

A comfortable breeze from the air conditioner, a steady 24 degrees Celsius, hit her in the face as she was greeted by the familiar burbling from the aeration tubes and hum of the air pumps. The slash-chink rhythm of Mario's machete, cutting the grass outside, was still audible over the automated systems of EVACC (pronounced like the first part of *evacuation*), the El Valle Amphibian Conservation Center. She washed her hands with antibacterial soap in the sink and put on a pair of powderless rubber gloves. At the top of a new page in the log, a cheap spiral notebook, she wrote: "21 December 2008, *Atelopus* room." Peering into the first couple of tanks, she found little to report in the log: "o fecal, o food left, one frog on plant, one hiding in peat moss." Spritzing each tank with a blast from the garden hose, with the nozzle turned to the mist setting, she quickly moved down the row.

EVACC is "a space age amphibian center nestled in the heart of an extinct volcanic crater" in the highlands of Panama, according to Lucy Cook, who writes the Amphibian Avenger blog. She describes EVACC as *"a terrifying vision of the future where frogs survive in sterile pods and crocs are mandatory footwear."*[1] The Amphibian Avenger blog is dedicated to "the ugly, the freakish and the unloved animals that are perilously ignored thanks to the tyranny of cute."[2] When Lucy Cook visited the EVACC facility, she was delighted to find "freakish" species, living alongside other frogs that might be conventionally considered cute. During my own visit to EVACC, I found human caretakers who had become emotionally and ethically entangled with creatures in their care, people who were committed to the practical labor of keeping frogs alive and flourishing in an era of extinction. Rather than terrifying visions of the future, I found practical applications of Donna Haraway's "cyborg politics," which involved forging new links between biotic elements and technology. Organisms and

FIGURE 4.1. Heidi Ross amid her daily routine of care at EVACC in El Valle, Panama. Photograph by Lucy Cooke.

machines had been joined together to ground modest hopes, creating the possibility for a shared future.[3]

Reaching into tank 12 with gloved hands, turning up each leaf, she struggled to locate all of the frogs. This tank was home to half a dozen lemur leaf frogs (*Agalychnis lemur*), which were decidedly cute.[4] Lemurs are nocturnal and spend daylight hours tucked up under leaves. At night, when the lemur leaf frogs are active, they are a reddish color, but during the day they assume a vibrant green. As she worked, some of the lemurs popped open their huge white eyes and began climbing toward the tank lid on spindly legs. The lemur tank was full of feces, as usual. She twisted the nozzle of the hose to jet and began washing the oblong blobs of mostly digested fruit flies through the wire mesh at the bottom of the tank.

In the wild, frogs hop away from their poop, reducing the risk of re-infecting themselves with diseases. Nematode worms and other parasites lay eggs in the digestive tracts of frogs and accumulate in the fecal pellets. Washing the poop away every day, and occasionally treating infected animals with drugs, helped maintain low parasite counts. Changing rubber gloves between tanks, or at least every time a glove contacted feces, protected the frogs from infecting each other. Every few days the unbleached

FIGURE 4.2. A robust population of lemur frogs (*Agalychnis lemur*), a notably "cute" frog species, lives in the EVACC facility. Until recently this species was thought to be extinct in the neighboring country of Costa Rica, but then breeding populations were found on an abandoned farm and in a forested area near Barbilla National Park. Photograph by Edgardo Griffith.

paper towels lining the bottom of the tanks were discarded and replaced. The trash stream of rubber gloves and paper towels was only one dimension of the costs incurred in keeping these endangered amphibians alive. Each month EVACC was running up an electric bill of around $800.

The EVACC facility is a fragile bubble of happiness sustained by a husband-and-wife team: Edgardo Griffith, a twenty-eight-year-old Panamanian biologist who often sports surfer's glasses, and Heidi Ross, an expatriate from the United States. It was created as a response to the growing sense of dread as the chytrid fungus spread across the highlands of Central America in a steady wave, about fifteen miles a year. In March 2006, Edgardo "spotted a dead frog in a stream near El Valle. Its limbs were splayed out, and its skin was peeling. He scooped it up, went home and cried."[5] As the disease hit, Edgardo and Heidi set about collecting frogs and keeping them in conditions of strict biosecurity. They began working as *bricoleurs* and entrepreneurs, assembling networks of organisms and objects, making do with beings and things that were ready at hand. Hundreds of frogs took up temporary residence in a few vacant rooms of Hotel Campestre, a backpacker hotel in El Valle. Edgardo and

FIGURE 4.3. The life cycle of the banded horned tree frog (*Hemiphractus fasciatus*) gives Rosalyn Diprose's notion of corporeal generosity a new twist. Eggs get pushed into a sack on the female's back as the male fertilizes them. These frogs breed by direct development, which means that when they are born, they pop out as small frogs instead of tadpoles. Then they stick around for a while, taking a ride on their mother's back. Photograph by Brian Gratwicke. Diprose, in her book *Corporeal Generosity*, contends that generosity is an openness to others that is critical to our existence, sociality, and social formation.

Heidi began to cobble together everyday technologies into a life support system to protect frogs from the pathogenic chytrid fungus.

In other parts of the world, the role of chytrid fungi in the mass extinction of frogs has been less clear. "Blaming the mass-extinction of amphibians on the chytrid fungus lets humans off the hook," said Deborah Pergolotti of the Frog Hospital in Cairns, Australia. Deborah told me about how a multiplicity of agents—like human real estate developers, ubiquitous chemical toxins, and diverse parasites—impinge on the precarious lives of frogs. In addition to chytrid infections, Deborah told me that the frogs of Australia were in decline as a result of tapeworms, viruses, and bacterial infections. One definitive review found that 233 amphibian species were in decline because of habitat loss or collection for the pet trade, while 207 species in intact ecosystems were experiencing "enigmatic declines" as a result of disease or climate change.[6] Conventional conserva-

FIGURE 4.4. Edgardo Griffith and Heidi Ross inside EVACC. Photograph by Lindsay Renick Mayer.

tion practices, strategies for protecting habitat, were no longer working with these enigmatic declines. Edgardo and Heidi were working at the vanguard, quickly developing new tactics and strategies as an emergent disease reconfigured local ecosystem dynamics.

Against the backdrop of mass death of loved animals, Edgardo and Heidi engaged in practical labor to maintain their optimism and hopes, helping maintain the world so that some good elements of life could continue.[7] Good feelings began to accumulate around these endangered frogs, to borrow the words of Sara Ahmed, such that these objects became "sticky." "Smiling, laughing, expressing optimism about what is possible will affect others," writes Ahmed. "Being with and around a person in a good mood gives a certain lightness, humor, and energy to shared spaces, which can make those spaces into happy objects, what we direct good feelings toward." Alongside their practical labor, Edgardo and Heidi also performed emotional and imaginative labor to sustain EVACC as a happy space where people could take pleasure in the well-being of animals with a magnetizing power of attachment. The happy feelings surrounding their facility and the amphibians in their care became contagious and started to affect others.[8]

One species in particular, the Panamanian golden frog (*Atelopus zeteki*), quickly became a poster child for international conservation efforts. In any absolute sense, the golden frog is not any cuter than the lemur frog or the horned tree frog. But as this species reportedly went extinct in the wild in 2008, Edgardo and Heidi became famous since they were keeping the last known populations alive in Panama within their facility. Golden frogs are featured on Panamanian lottery tickets and have been scripted into stories about national patrimony and heritage. Locally the couple became renowned for their golden frog mobile, a four-wheel-drive Jeep painted yellow with black stripes. In conservation circles, *Atelopus zeteki* quickly became a flagship species—a surrogate that began to stand in for other unloved frogs in fund-raising campaigns. These charismatic animals captivated the imaginations of professional conservation biologists, volunteers, and donors, because they fit a new hopeful story line. On the brink of extinction within its natal ecological communities, the golden frog was saved by technological and scientific interventions.

Amid a demanding routine of daily care for the frogs living in Hotel Campestre, Edgardo and Heidi began to navigate oblique powers structuring uneasy North–South relations.[9] Wrangling with diverging values and obligations, they explored nonhierarchical modes of coexistence with large institutions. Major donors from North America—namely the Atlanta Botanical Gardens, Zoo Atlanta, and the Houston Zoo—began to lay the foundations for the EVACC buildings nearby, tailor-making a biosecure facility at a local zoo to replace their makeshift facility at Hotel Campestre. The Amphibian Ark, the transnational organization I learned about in the Reverted Zone, became involved only after Edgardo and Heidi moved their frogs to the new building. With a mission to "ensure the global survival of amphibians," the Amphibian Ark was drawing facilities like EVACC into a global network of institutions "focusing on [species] that cannot currently be safeguarded in nature."[10] The Ark was producing hope for endangered frogs at the intersection of concrete actions of care in the present and messianic dreams about a future to come.

The Amphibian Ark dubbed 2008, the year I visited the EVACC facility, as the Year of the Frog. Kevin Zippel, the founder and principal visionary of this global ark, was aiming to raise a $50 million endowment to preserve endangered frog species in perpetuity. In October 2007, as he prepared to launch the Year of the Frog, Zippel gave a speech at the Jackson Hole Wildlife Film Festival, where key players with National Geographic, Smithsonian, Discovery, Animal Planet, *Nature*, and several other international media organizations were in attendance. In his speech, where he showcased the plight of the golden frog, he opened with a provocative line: "Hi, I'm Kevin, and I'm building an ark." Zippel continued:

> Amphibians are our modern-day canaries in the coal mine. Just as the miners would take these sensitive birds with them into the mines, and they would know if the birds died it was time to get out, amphibians are raising and waving red flags to us, saying, "There is a serious problem, you need to change your behavior, or you are going to suffer the same consequences."
>
> A recent assessment of all amphibian species revealed that nearly half are declining. Somewhere between a third and a half are threatened with extinction. Just within in the past few decades well over a hundred have already gone extinct. This is far more severe than what we see with other vertebrate groups. And for every bird or mammal species that is threatened with extinction, there are two to three amphibian species that are on the verge.[11]

Kevin's rhetoric echoes Al Gore's language in *An Inconvenient Truth*, a film about global warming. This film has a secular apocalyptic narrative, with a revoicing of environmental science in the language of evangelical Christianity, according to Susan Friend Harding. These narratives are jeremiads, a type of Protestant political sermon lamenting that people have fallen into sinful ways and face ruin unless they swiftly reform. "Doom is imminent—but conditional, not inevitable. It can be reversed by human action, but time is short."[12] As canaries in the coal mine, amphibians stand in as vulnerable surrogates for humans. Ongoing extinctions of frogs, salamanders, and caecilians prefigure a possible future event in Kevin's imagination—the extinction of the human species.

As a figurehead at the helm of the Amphibian Ark, Kevin Zippel was not just focused on definitive endings, the apocalyptic thinking critiqued by Derrida and many other thinkers. Zippel was working to embody the messianic spirit. As scores of frog species were going extinct, he was trying to open up a moment of revolutionary time—a moment when collective hopes about "saving the environment" or "preserving nature" might coalesce around the future of actual animals.[13] Derrida suggests that we should embrace the messianic by literally expecting the unexpected, by waiting for mysterious possibilities that are beyond our imaginative horizons. Rather than pinning hopes on something concrete, Derrida would have us wait for nothing in particular.[14] In contrast, Zippel's imagination was focused on something specific: creating a livable future for a multitude of endangered animals.

Kevin Zippel was grounding modest biocultural hopes in hybrid assemblages of nature, culture, and technology.[15] Given the scale of the extinction wave sweeping through the worlds of amphibians, Zippel recognized that isolated local efforts, like the EVACC facility in the highlands of Panama, were not capable of addressing the global crisis at hand.[16] He was doing the difficult imaginative and emotional labor to sustain hope in a seemingly hopeless situation. I first watched a YouTube video of Zippel's canary in the coal mine speech in 2008, with Edgardo and Heidi. They told me that Zippel was helping them hold on to hope and happiness, as they went through mind-numbing daily routines to care for the thousands of animals in their facility. Against a prevailing sense of homogeneous, empty time—when nothing really seemed to change even amid definitive extinction events—Kevin Zippel was trying to open up revolutionary possibilities with his messianic language.[17]

Working to galvanize the conservation community, Zippel hoped to raise an endowment to preserve frogs for eternity in biosecure breeding facilities and cryogenic banks. He was trying to show others the magnetizing power of amphibians, to generate contagious feelings of happiness around efforts to preserve their endurance, well-being, and survival.[18] Happiness, however, is like a fragile bubble. It can shatter, like glass, at any moment.[19] The Year of the Frog, 2008, coincided with global financial disaster. Failing to raise even a fraction of their ambitious goal, raising less than $1 million out of the $50 million goal of their capital campaign, the Amphibian Ark barely stayed afloat, barely covered their 2008 operating expenses.

LIVE FREE OR DIE

Frogs who experience the ambivalent grace of salvation probably do not share the subjective feelings of happiness sometimes experienced by their human caretakers. Frogs have very different faces from humans. Our smiles, big toothy grins, likely look very different to these small vulnerable animals when we lovingly grasp them in our hands. Sara Ahmed notes that many people experience "inappropriate affects" that are out of sync with social expectations. The figure of the unhappy bride on her own wedding day is one of the examples studied by Ahmed. "We become strangers, or affect aliens, in such moments," Ahmed writes.[20] Endangered animals are probably often alienated from human affects. While zoos and other breeding facilities style themselves as "salvific arks, bearing life's remnant and our hopes for redemption," this rhetoric often masks widespread feelings of unhappiness and even hidden regimes of violence.[21] As the messianic vision of Kevin Zippel largely failed to materialize—in the absence of an endowment capable of sustaining the life of endangered frog species in perpetuity—many committed caretakers like Edgardo Griffith and Heidi Ross soldiered on with limited resources, working to imagine and craft better futures for the animals in their facility. Others, with different political and ethical commitments, adopted more violent forms of care.[22]

After returning to the United States, I learned that a breeding population of golden frogs had been airlifted out of Panama in 1999, the same year the U.S. military occupation was coming to an end. Scores of golden frogs were collected for a captive breeding program in the United States aimed at "conserving genetic variability and maintaining viable captive populations." The Maryland Zoo in Baltimore was given an import permit, in accordance with the Convention for the International Trade in Endangered Species (CITES). This permit granted "ownership of the animals" to the zoo, rather than the Republic of Panama.[23] After a few false starts, the Maryland Zoo had much success in breeding golden frogs. Perhaps they were too successful. Every time breeding pairs mated they produced some two hundred to nine hundred white eggs. Initially the biologists overseeing this conservation program were delighted at the prolific nature of these animals. They were happy to see them flourishing within artificial ecosystems. Soon, however, the Maryland Zoo ran out of space. They began shipping frogs around the United States—in plastic Gladware deli cups lined with damp toilet paper—to other zoos. These frogs are now common features of reptile houses. They are on display at institutions throughout North America, like the Bronx Zoo, the Smithsonian's National Zoological Park in Washington, DC, the Atlanta Botanical Garden, the Toronto Zoo, and Busch Gardens in Tampa, Florida.[24]

As the zoological community began to run out of space, zookeepers started killing Panamanian golden frogs by the hundreds. Even after this species was presumed extinct in the wild, zoos culled captive populations—selecting the most "genetically valuable" individuals to live. At Herp Happy Hour in Washington, DC, a monthly meetup where herpetologists (reptile and amphibian experts) go for hard drinking, I met some insiders in this program. One zookeeper told me: "Every time we have a new clutch of golden frogs I have to select sixty of the healthiest frogs to live. I can't stand the job of killing an endangered species, so I make my boss come in and euthanize the ones I don't select."[25] Writing of related dilemmas among bird conservationists, Thom van Dooren suggests, "Many of us would still choose the violence of a conservation grounded in captive breeding over that of extinction." Even still, van Dooren insists that we consciously dwell with the genuine ethical difficulties "in an effort to, wherever possible, work toward something better."[26]

Writing of interspecies love in the age of extinction, Deborah Bird Rose argues for an ethics care that does not exclude death. "An ethical response to the call of others does not hinge on killing or not killing," she argues. The question becomes, what constitutes a good death? One prominent frog biologist who was also at the Herp Happy Hour in Washington told me that a "good death" cannot come from euthanasia at the hands of a zookeeper. Amid a sedate and melancholic conversation about biodiversity loss, financial woes, and zoo overcrowding, she suddenly slammed down her glass, spilling margarita on the table. Lifting her hand in a parody of a revolutionary salute, she shouted, "Live free or die!"

The deadly chytrid fungus is still present in the highlands of Panama. If the golden frogs were reintroduced to Panama and released, most would probably die. But by the reckoning of the researcher who spilled her margarita, a good death in the wild connected to the hopes of adaptation and survival is better than a bad death at the hands of a zookeeper. Despite hopes that some robust frogs might live if released in the wilds of Panama, influential members of the conservation community are reluctant to let them go. Zoos are cosmopolitan collections of animals, breeding grounds for diseases from diverse corners of the world. Michel Foucault understood the modern zoological garden as "a sort of happy, universalizing heterotopia" where "several spaces, several sites that are in themselves incompatible" are juxtaposed in a single real space.[27] Findings by veterinary pathologists suggest that zoos might be better understood as heterotopic hotbeds of parasitic protozoa, fungi, viruses, and bacteria. The officials of the Maryland Zoo are thus reluctant to send thousands of golden frogs back to Panama when they might inadvertently unleash a new amphibian disease from U.S. zoological collections into Central America.

THE FROG FRIDGE

Rather than just critiquing standard zoological practices for managing endangered forms of life, I began volunteering my time to care for frogs that had gone extinct in the wild. As a participant observer at facilities associated with the Amphibian Ark in Panama as well as the Bronx Zoo in New York City, I cleaned cages, prepared food, and fed animals alongside zookeepers who were overworked and underpaid. While observing the toll of monotonous routines on human laborers, and the cramped conditions for tens of thousands of animals living with regimens of institutionalized care, I crafted a concrete proposal for doing things differently. In collaboration with Grayson Earle, a digital artist, and Mike Khadavi, a frog enthusiast who designs custom aquariums, I created an artwork—*The Utopia for the Golden Frog of Panama*—in hopes of saving a few animals from euthanasia. It was my attempt to dwell with the genuine ethical difficulties of keeping endangered animals in captivity and work toward something better.[28]

Utopian worlds are ever present in science fiction, where the projection of new heavens is never far from the emergence of new hells.[29] Utopias can also function as diagnostic tools. To paraphrase Isabelle Stengers, they are learning grounds for resisting what today opportunistically frames our world.[30] *The Utopia for the Golden Frog* was built out of an unused refrigerator enhanced with custom digital equipment, an aquarium, and a living ecosystem. This installation was my best attempt to interpret the interests and needs of another species. We adapted and used technologies that were ready at hand: household appliances, cheap digital hardware, and some specialized equipment from pet stores.[31] Hacking into the refrigerator with a power saw, we put a glass window in the front door. Grayson Earle also hacked into the electrical system of the refrigerator, creating a digital thermostat, using an Arduino, a small programmable microcontroller, to keep the fridge within 68–73° F daily, the ideal range preferred by *Atelopus zeteki*.

The Arduino microcontroller also regulated the power of the refrigerator and controlled our tailor-made system of heat lamps. An automatic timer kept the lamps on a twenty-four-hour light-dark cycle. If the fridge became too hot, the cooling system was activated. The system also had an emergency kill switch for the lights, cutting them off if the cooling system failed to kick in. This portable amphibian pod had all the technical and architectural components that would let frogs from the cool highlands of Panama live through a hot New York City summer.

Many frogs in facilities associated with the Amphibian Ark live in sterile tanks, sitting day after day on a damp paper towel. Perhaps they experience a sense of cosmic loneliness—isolated from other species and companions that make forests livable and lively places. As an antidote to this potential loneliness, my other collaborator, Mike Khadavi, assembled a miniature ecosystem inside the refrigerator. Useful mosses and vascular plants collected from diverse corners of the globe were installed in the fridge to generate enough oxygen to keep a small population of frogs

FIGURE 4.5. Grayson Earle pictured next to *The Utopia for the Golden Frog* at *The Multispecies Salon* in Brooklyn. Photograph by Eben Kirksey.

alive. This feature, which meant that this frog fridge rarely needed to be opened, arguably made it more biosecure than the large pods of the Amphibian Ark, where humans were constantly coming and going.

While we made no pretense of establishing conditions of sterility, conditions that are no more achieved in zoos or other Amphibian Ark facilities, we carefully selected other species that are good for frogs to live with in multispecies worlds. After ordering mutant fruit flies from an online retailer (Ed's Fly Meat, www.flymeat.com), we created a special composting system and seeded the tank with wingless *Drosophila melanogaster* and flightless *Drosophila hydei*.[32] Aside from the occasional addition of human food waste to the composting system, to generate future generations of fruit flies, *The Utopia for the Golden Frog* was built to function in relative autonomy, as long as it was plugged into an electrical outlet.

The title, *Utopia for the Golden Frog*, was pregnant with irony. Living in conditions of incarceration is certainly not utopian. Rather than communicating a fixed vision of how things should be, it was meant to illustrate possible biocultural articulations. This fridge for frogs was a cousin of the Fridgeezoo, a small plastic toy described by Johanna Radin and Emma Kowal as "an avatar of anxiety" in an era of planetary warming.

The Fridgeezoo toys are shaped like one of four charismatic species—the walrus, polar bear, penguin, or seal—and have a light sensor that triggers a recorded greeting, "Hello!" when the refrigerator door is opened. But this simulated Arctic refuge becomes agitated if one lingers indecisively, yelling, "Close the door! You are wasting energy!" Radin and Kowal use the Fridgeezoo to introduce their collaborative study of cryopolitics—the efforts to produce cold temperatures as a means of delaying processes of decay. They apply the notion of cryopolitics to understand how organisms, tissues, and species are preserved in suspended animation with a deferred death.[33]

Cryopolitics joins Michel Foucault's influential notion of biopolitics—the political art of making certain people or things live, while letting others die. Our biosecure holding tank drew inspiration from *Tactical Biopolitics*, a book by Beatriz da Costa and Katia Philip that brings the ideas of Foucault into conversation with tactical media practices in the arts. If Foucault understood biopolitics as centralized forms of optimization, surveillance, and control, then bioartists have begun to use surprising tactics to expose and derail dominant practices for managing life.[34] Following this artistic tradition, we mounted a webcam inside the frog fridge for live viewing, and posted a digital archive of temperature and humidity readings on the Internet, enabling anyone who was interested to verify we had met the technical requirements to sustain the life of this species. In other words, we exposed our artistic intervention to dominant regimes of biopolitics—opening up the frog fridge to monitoring by the zoological community or surveillance by government authorities.[35]

After assembling *The Utopia for the Golden Frog* and installing it for an upcoming gallery opening—*The Multispecies Salon* at Proteus Gowanus in Brooklyn—I sent an e-mail to an employee at the Maryland Zoo in Baltimore asking after some frogs.[36] The person I e-mailed was the "stud book holder" for the species, a designation by the Association of Zoos and Aquariums (AZA) for the person who "dynamically documents the pedigree and entire demographic history of each individual in a population of a species."[37] My e-mail asked about how "I might submit a formal application to borrow some *Atelopus zeteki* adults for a temporary artistic display about the amphibian mass extinction crisis." I outlined the technical specifications of the frog fridge, adding, "The tank does not have any running water, so based on what I have learned about *Atelopus* reproductive biology I trust that this means the frogs won't be trying to breed."

The stud book holder wrote a friendly but dismissive note back, saying that he was unable to provide golden frogs for exhibition at facilities that were not zoos, institutions licensed by the AZA. The original permit from the U.S. Fish and Wildlife Service specifically prevented the zoo from giving any members of this species to private individuals. Golden frogs are listed under the U.S. Endangered Species Act, which means that they are very heavily regulated, and their movements are governed by many layers of paperwork. Any proposal for research or exhibition outside AZA networks would need to clear the Maryland Zoo's Institutional Animal Care and also be authorized by U.S. Fish and Wildlife. In parallel to my e-mail exchange with the stud book holder, I lobbied influential movers and shakers in amphibian worlds. But I ultimately failed to convince the Maryland Zoo and the U.S. Fish and Wildlife Service that a few frogs should be saved from euthanasia and kept in our modified refrigerator.

The Utopia for the Golden Frog failed to house an actual animal, but perhaps it still succeeded as an artwork—functioning as a critique of dominant practices for managing life and death. To again borrow the words of Isabelle Stengers, the installation served as "a diagnostic vector for what made it a mere utopia, as a learning ground for resisting what today opportunistically frames our world."[38] Calling up Kevin Zippel, founder of the Amphibian Ark, I used this artwork as a para-ethnographic object, a conversation piece, to facilitate unconventional ways of speaking and thinking about the issues at hand.[39] Sharing a link to the website for the frog fridge with Zippel, I told him that it was a modest proposal that might let citizen scientists participate in the effort to care for endangered species. Revoicing his own words, I said, "It seems like any way of expanding the carrying capacity of the Ark would be a good thing." "I agree," Zippel responded. "But, a lot of people in the private sector are motivated by the value of these animals. There is a tendency to sell them on the black market. There are just so many complicating factors to citizen involvement."

People who trade in endangered species, who collect them in the wild and sell them as pets, are one of the problems leading to extinctions. The pet trade for reptiles and amphibians is a booming informal economy in the United States. As "owners" of the golden frogs, the Maryland Zoo was entrusted by U.S. Fish and Wildlife to guard against their escape into the market for rare and exotic animals. While they were not overly concerned with the fate of the animals with little "genetic value," the imagined excess of their breeding programs, ultimately their goal was to protect the viability of these frogs in the wild. A small population of golden frogs was recently rediscovered in Panama. Hopes of conservationists, who once presumed that this species was extinct outside of captive breeding facilities, have been placed on the living figures of these animals. The fear of the conservation officials, Zippel intimated, was that people would collect any remaining golden frogs in Panama, some of the most "genetically valuable" individuals in existence, and then launder them as animals bred in captivity.

Protecting frogs from market forces, prizing the genetic value of animals over their potential exchange value, has some merit. Turning animals into commodities for exchange certainly has the potential to bring about troubling regimes of exploitation. But, against the backdrop of a grim political and economic situation, leveraging the economic value of these animals might be the best way to actualize Kevin Zippel's conservation goals. Only fifty species found a home in his Ark, living at institutions that might be able to sustain them over the long term.[40] This meant that there were some 3,850 species of frogs with declining populations beyond the Ark's carrying capacity. Zippel's Amphibian Ark was fragile, needy of care. Many keepers working in facilities associated with the Ark told me that they struggled to keep the faith as animals inexplicably died, or as entire species went extinct despite the best efforts of the conservation community.

Optimism attached to other living beings, according to Lauren Berlant, can be the cruelest of all. When contingencies beyond one's control result in the death of a loved one, this produces a cruel, jarring slap.[41] Maintaining optimism, as loved animals or entire species die, involves difficult emotional and imaginative labor. Converting despair to hope involves playing with the uneasy alchemy of the pharmakon, turning obstacles into opportunities, transforming poison into a cure.[42] Converting from indifference—"it is too late, the world is shit"—to a more hopeful orientation is difficult. "Having someone to care for, and thus caring for what happens, caring about whether there is a future or not," makes it easier to sustain happiness, according to Ahmed. Hope thus involves vulnerability when you "care for that which is beyond or outside your control." Hope, in other words, involves caring about "the *hap* of what happens."[43]

"Having good 'hap' or fortune," notes Ahmed, was the original sense of the word "happy" in Middle English. While this meaning may now seem archaic—since happiness is not something that money can buy or power can command—Ahmed insists that we return to this original definition "as it refocuses our attention on the 'worldly' question of happenings."[44] Efforts to postpone the processes of extinction, to hold species in the suspended animation of cryopolitics, seek to control the hap of what happens. Freezing the genetic diversity of living plants or animals within

an ark, and managing them according to carefully calculated biopolitical rules, risks producing and sustaining forms of life that are no longer
happy. Managed or directed life does not have hap, as in happenstance
opportunity.[45] While holding listless animals inside life support technologies, many conservation biologists are only able to hold on to happiness
by imagining the moment when they are given license to break open their
fragile bubbles of glass. Their emotional and imaginative labor involves
picturing the escape of animals they love from their present circumstances. They imagine a happy future for their loved ones in the wild,
beyond regimes of care and control.

INTO THE WILD

Wildness once existed only beyond the reach of civilization and domestication in the popular imagination of Europe.[46] On American frontiers,
wildness has long been regarded as something to be tamed, subjugated,
and brought under control.[47] Pushing past visions of colonialism and empire, Sarah Franklin has described wild forms of life emerging within the
domains of biotechnology. "The new wild is the successor condition of
the old wild," writes Franklin, "which meant nondomesticated, as in 'wild
geese' or 'a wild boar.' What might be described as overdomestication or
hypercultivation turns out now . . . to produce the other sense of 'wild,'
as in something dangerous, risky, and out of control—'a wild idea' or 'a
wild night out.'"[48]

Casual visitors to Amphibian Ark facilities who are technophiles often
comment on its wildness in the sense described by the Urban Dictionary:
"out of control, out of this world, an unrestrainable amount of coolness."[49]
Others see risky and dangerous elements in this biotech assemblage that
includes mutant fruit flies and biosecurity protocol.[50] After many months
of participant observation at facilities associated with this Ark, I became
less concerned about possible risks inherent in the project and more
interested in the sense of wildness at play in the imagination of many
people who care deeply about happy futures for frogs. Rosemary-Claire
Collard asserts that the notion of the wild can be rehabilitated from the
discourse of colonialism to recognize the autonomy and alterity of other
species. The phrase "wild life," according to Collard, might refer "to an un-
captive life that retains the ability to work for itself and others in its social
and ecological networks (potentially including humans)." Wild animals,
she asserts, have the capacity for bodily flight as well as the freedom to
raise families of their own.[51]

In November 2013, curators and keepers actively working with Panamanian golden frogs traveled from the United States to El Valle, the site of Edgardo and Heidi's EVACC facility, to discuss plans about releasing captive-bred frogs back into the wild. Twenty-seven research scientists, conservationists, zookeepers, and government officials were involved. They worked through tangled thickets of ethical, logistical, political, and epidemiological issues that plague conservation initiatives across national borders. Passions became inflamed, according to one attendee, when the subject of repatriating the golden frogs currently living in the United States was discussed. No one brought up the uncomfortable subject of euthanizing frogs, nor the long legacy of unilateral action by U.S. agents in the region, but the Panamanian delegation did repeatedly insist that their North American counterparts urgently needed to start finding new creative solutions to the problems at hand.[52]

The U.S. zoological community had enfolded the golden frog into an emergent ecosystem—a system of holding tanks, public displays, and revenue-generating visitor attractions—that showed promise of sustaining itself into the future. But the frogs that had become caught in this complex assemblage were being held fast by agents and institutions with competing values and obligations. In a phrase, Panamanian golden frogs had become entangled in relations of "reciprocal capture." "In the case of symbiosis," writes Isabelle Stengers, reciprocal capture "is found to be positive: each of the beings coinvented by the relationship has an interest . . . in seeing the other maintain its existence."[53] The case of golden frogs and U.S. zoos is an example of reciprocal capture where the relationship was not entirely positive. U.S. zoos were making thousands of individual animals live in uncomfortable situations, while making other animals die that did not fit within their system.[54] In my own mind, I could only find hope in this contingent relationship if it contained the possibility that the frogs might one day escape—returning to the wilds of Panama.

Vicky Poole, a dedicated zookeeper who has been a key player in Project Golden Frog for over a decade, told me that everyone at the El Valle summit shared the same ultimate goal of sending some frogs from the United States back to Panama. But government regulations "took all of the fun" out of the process, in Vicky's words. Generally, U.S. Fish and Wildlife permits to export endangered species involve a two- to three-year process. Precautionary principles loomed large in my conversation with Vicky—she told me that the U.S. conservation community was reluctant to send frogs back for fear of introducing a new unknown disease, since suitable habitat for the frogs was being destroyed as "Panama is developing at such a fast rate now." When we talked, in September 2014, Vicky and her colleague Kevin Barrett, the stud book holder at the Maryland Zoo, had not yet submitted an export permit to the U.S. government. Even after this export permit is submitted and approved, they plan to continue euthanizing frogs. "Since we will only be sending bloodlines down for breeding in Panama," Barrett told me, "we won't need to send thousands of frogs down." Vicky Poole added that "chasing the paper" with U.S. Fish and Wildlife made it "logistically difficult" to send large numbers of frogs to Panama.[55]

Allowing for wilder dynamics with more permissive legal frameworks might give endangered frogs more opportunities for good hap or fortune. Captive animals would have the potential to rediscover happiness if they were given more opportunities to escape beyond the circumscribed visions of the future maintained by the zoological community. Perhaps some of these literal amphibians have the capacity to break through to ontological cagelessness, to go wild along unpredictable lines of flight—to find homes in new worlds even if their previous worlds have been destroyed or altered beyond repair. Ecosystems now embody elements of the old wild as well as the new wild. Since emergent ecological communities do not look like past environments, visionaries are starting to imagine how new biological companions and technical rivets might help anchor frogs in the world. They are imagining new ways to support the autonomy and alterity of endangered animals by fostering new relationships among companion species.[56]

FUTURE PROMISE

While conservation practitioners crafted piecemeal solutions to deal with the fact that many frogs can no longer live without protective infrastructures, some scientists were working at the frontiers of their imaginative horizons, searching for a breakthrough cure for the chytrid fungus. The Smithsonian National Zoo announced in 2012 that it was inoculating frogs with experimental probiotic treatments in Front Royal, Virginia. "We usually think of bacteria as bad for us, but that isn't always the case," heralded the Smithsonian website. "For us humans, the most common examples of helpful bacteria, or probiotics, live in yogurt." The blog described how the redback salamander, a native of the eastern United States, had survived the epidemic chytrid outbreak. These salamanders had a diverse array of microbes living on their skin. Researchers speculated that these multispecies communities could be a probiotic shield guarding against infection—a thin living bubble of protection. They hoped to discover microbes that could become new companions for endangered frogs, capable of producing chemical compounds with antibiotic (or at least antifungal) properties.[57]

Hopes were placed on one kind of bacteria, *Janthinobacterium lividum*, that produced an antifungal compound called violacein in the lively microbiome on the skin of redback salamanders. When transferred to the mountain yellow-legged frog, an endangered species from California, these bacteria were protective against chytrid fungal infections. Captive Panamanian golden frogs, dubbed by the blog as "the poster-child for amphibian conservation," were also inoculated with probiotic bacteria treatments along with pathogenic chytrid fungi in experimental trials. While *Janthinobacterium lividum* bacteria initially kept fungal infections on golden frogs to a low level, eventually the probiotic microbes decreased in abundance and the frogs died.[58] In the face of this failure, I found cautious hopes proliferating in scientific dreamworlds. Searching through microbial worlds, I found hope coalescing around specific figures, only to quickly dance away again on other lines of flight. In Panama I talked to researchers who were carefully testing some six hundred kinds of microbes from the skin of other frogs—living figures of hope that might one day enable a multitude of golden frogs to again live in the wild.[59]

Speculation about scientific breakthroughs often fuels messianic thought in the biosciences.[60] "There can be no science without speculation," writes Mike Fortun, "there can be no economy without hype, there can be no 'now' without a contingent, promised, spectral and speculated future." If messianic speculation in biology is often articulated to moneymaking dreams and schemes, different political, economic, and ethical forces are at play in research initiatives aimed at curing amphibians.[61] A probiotic treatment for endangered frogs, which pins hopes on the figure of microbes, would not likely result in the payout of actual capital, as is the case with blockbuster pharmaceutical drugs for humans.[62] Instead, a cure would just generate symbolic capital for researchers within university communities.[63] A scientific breakthrough involving these symbiotic microbes could nonetheless actualize the vision of salvation promoted by the Amphibian Ark. Caretakers who maintain this fragile ark are laboring in the present, working with scarce resources and following mind-numbing routines in the now, while harboring dreams of an unknown future to come.

Concrete hopes are emerging with these cosmopolitical dreams. Researchers and amphibian caretakers are harboring fantasies of forming new political articulations with microbial allies, enfolding them into a stable imagined world against the backdrop of an unknown and potentially unstable cosmos. Departing from these dreamworlds, I began to ramble through microcosms, focusing my attention on critters that are just at the edge of human bubbles of perception and action. Following chytrid fungi through scientific networks, I began to consider the happy accidents and chance encounters that enable microbes to exploit faults and fissures within established assemblages. While studying an emergent disease, a pathogenic chytrid that had generated dramatic shifts in conservation strategies and new speculative trajectories in the biological sciences, I let my attention wander. While sifting through the flotsam and jetsam of the microbial cosmos, I found other fungal species that were invisibly working to sustain my own existence within ecological networks.

CHAPTER FIVE

BUBBLES

. . .

Peter Sloterdijk describes worlds as bubbles. In the first volume of his magnum opus, *Spheres: Bubbles, Microsphereology*, he asserts, "Spheres are by definition morpho-immunological constructs. Only in immune structures that form interiors can humans continue their generational processes and advance their individuations."[1] Certainly the figure of the bubble has much generative potential. But Sloterdijk has made some fundamental errors at the foundations of his sphereology. He asserts that humans are exceptional bubble makers. "Wherever human life is found," writes Sloterdijk, "whether nomadic or settled, inhabited orbs appear, wandering or stationary orbs."[2] Any serious student of natural and cultural history knows that we, as humans, are not exceptional. Anyone who is truly worldly knows that we share planet earth with multiple species of ontological amphibians.

Chytrids are unloved microbes. They live all around you, beyond the purview of your everyday awareness. A swarming multitude is constantly invading and destroying the spheres of immunology, and also making them.[3] Chytrids are liminal critters that trouble our categories. Most microbiologists call them fungi. But seemingly stable formations like "fungi," "the animal," or "species" fall apart when you look too close.[4] When chytrids are young, in the zoosporic stage, they resemble human sperm and constantly swim about. Once they find a suitable substrate, like a grain of pollen or the skin of a vertebrate host, they put down root-like structures called rhizoids. When isolated in pure culture, chytrids usually have stable morphologies as they grow. Chytrids generate spheres nested within other spheres, clear bubbles containing darker bubbles. When surrounded by other beings and things, when living in microbial ecosystems, chytrids are often ontologically indeterminate. They often become something else, adopting completely different structures, depending on who or what is in their world.[5]

One species of chytrid, *Batrachochytrium dendrobatidis*, is destroying the worlds of amphibians. In mid-August 2012 I followed this microbe to a laboratory at the University of Maine run by Joyce Longcore, the lead author on the paper giving this pathogenic microbe a name. Volunteering my time, while taking copious notes, I helped care for a collection of several hundred kinds of chytrids. Creating and seeding microcosms, I helped contain the chytrids in bubbles, common worlds shared with humans. "One of my pet peeves," Joyce confided shortly after I met her, "is how people talk about 'chytrid fungi.' The way people talk implies that all chytrids are the same, that they are all responsible for the mass mortality of frogs." Many chytrid species perform critical ecological functions—some break down chitin, the hard material in the exoskeleton of insects, while others that live in the hind guts of ruminants help digest cellulose, a sugar molecule in dead plant matter that is hard to break down.

In preparation for my visit to Maine, Joyce asked me to bring water and soil samples from a favorite site so that I could find out what chytrids have been missing from what she called my "umwelt," or known world. Joyce was speaking my language. Umwelten, the phenomenological bubbles first described by Jacob von Uexküll, denote a ringed or enclosed world.[6] Humans are not that different from other kinds of animals, by von Uexküll's reckoning, in that we are also enclosed within a phenomenological sphere, a world of perception and action.[7] Peter Sloterdijk, who styles himself as the second coming of Martin Heidegger, uses this concept differently. He understands the umwelt as an ontological cage. Sloterdijk asserts that only animals are confined to umwelten, while human beings have experienced a "breakthrough in ontological cagelessness" that can be best characterized as "world."[8] Rather than bothering with Heidegger or Sloterdijk, Joyce Longcore's understanding of umwelt stays true to Jacob von Uexküll's original ideas. When I visited her lab, she illustrated the limits of my own ontological cage, and then patiently showed me how to tentatively grasp new species of beings. She showed me how to expand the reach of my own umwelt.

Humans need a microscope, a sensory prosthetic, to see chytrids. When new samples of soil and water first enter the Longcore lab, the first step is to hunt for chytrids in "gross culture"—in microcosms that are teeming with bacteria, other kinds of fungi, and monstrous animalcules. Joyce helped me find chytrids amid microbial flotsam and jetsam.

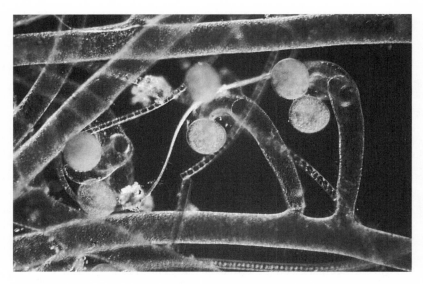

FIGURE 5.1. It is easy to become distracted while hunting for chytrids amid nematodes, rotifers, filamentous fungi, and debris. This image depicts yellow-green *Vaucheria* algae with zygotes. Photograph by Tom Adams / CORBIS.

As rotifers whirled in tight circles and giant nematodes poked around, I became easily distracted. "You have to look until your eyes start filtering out everything else," Joyce says. "Sometimes I crawl into the microscope," she adds. "I forget to breathe. I forget what I'm working on."

At an early age, Joyce's imagination was captured by chytrids. She was co-opted by these microbial strangers, enfolded into their worlds.[9] After receiving an undergraduate degree in biology at the University of Michigan in 1960, she worked as a research assistant for Professor F. K. Sparrow, an eminent student of zoosporic fungi. Joyce recently encountered a letter written by her twenty-one-year-old self, addressed to Sparrow. Reading between the lines of her own words, she found unbridled enthusiasm and a determination to devote her professional life to the study of chytrid fungi. But her passion for studying chytrids was put on hold by love for another human. Marriage to a U.S. Fish and Wildlife field biologist—a researcher who made key findings about DDT and thin bird shells—meant that her professional vision was postponed for nearly thirty years. After raising a family, she went back to graduate school and earned her PhD in 1991, at the age of fifty-two. She has never been the lead investigator of a major grant for her research, and is not on the uni-

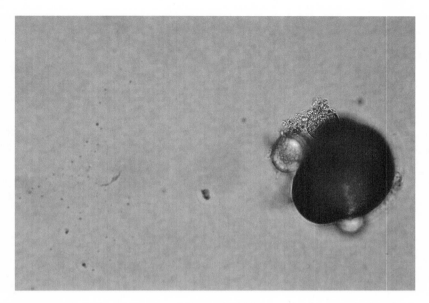

FIGURE 5.2. Chytrids appear as translucent bubbles attached to a dark globule of pollen. Photograph and microscopy by Eben Kirksey in the Longcore lab.

versity payroll, but after buying and borrowing equipment, she started her own lab at the University of Maine. Joyce has published dozens of papers, some of which are widely cited, and is internationally recognized as one of three experts on the taxonomy of chytrid fungi—a phylum comprising over one thousand species.

Joyce guided me with quiet enthusiasm as I searched through the microbial morass under one of her microscopes. She helped me spot a clump of chytrids anchored to a grain of spruce pollen, a source of food that she had placed in the gross culture as bait. Other common baits used by Joyce include chitin from shrimp exoskeletons, hemp seeds, and onion skin. Spruce pollen grains are preferred by human chytrid hunters over pine pollen, because of their larger size. The chytrids on the pollen grain were swollen in translucent fat spheres called zoosporangia. This structure is characteristic of all chytrids in their final stage of asexual reproduction. Dark rhizoids, thin rootlike filaments, branched out from the base of the zoosporangia, burrowing into the pollen.

Mature zoosporangia are clear bubbles that contain dozens of little darker bubbles that teem together in a tightly packed mass. As I watched, the black orbs began to vibrate and wiggle inside one zoosporangium (the

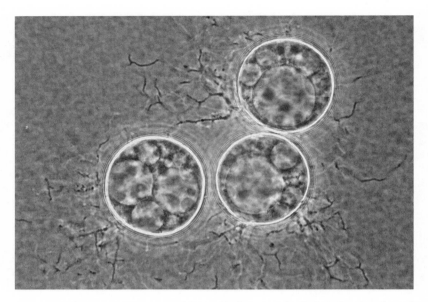

FIGURE 5.3. Three zoosporangia of *Batrachochytrium dendrobatidis*, the chytrid that is destroying the worlds of amphibians. The dark rootlike rhizoids burrow into the growth medium, while translucent spheres inside of these spheres grow into new zoospores. Photograph and microscopy by Joyce Longcore.

singular of zoosporangia). Suddenly the dark balls, baby chytrids or zoospores, began shooting out of an opening emerging from the side of their parents. These sperm-like critters have long whip-tailed flagella. Within a minute or two the zoosporangium emptied out. A few zoospores were left, thrashing about inside. Soon these too escaped. Before my eyes, the zoosporangium became an empty husk. Adult chytrids are single-celled bodies that have the usual suite of organelles shared by eukaryotes (animals, plants, and fungi).[10] Certainly they inhabit umwelten of their own. Dorion Sagan has described a bacterium swimming up a sugar gradient as a living being that is "aware of the signs of its own continued being and thus contrarily of its own potential demise."[11] Chytrids are also certainly purposeful beings with an awareness of the presence and absence of nutrient sources. They also have an ability to grasp other beings who share common worlds.

Gazing into a microscope, it would be easy to follow Sloterdijk in assuming that only humans look out into horizons, gazing into the mysterious beyond outside of established worlds. But as I watched zoospores

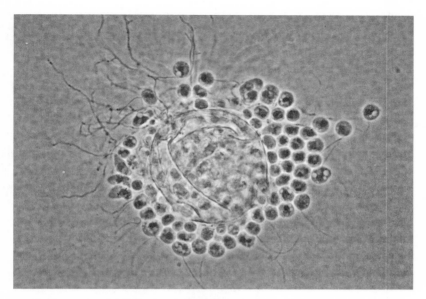

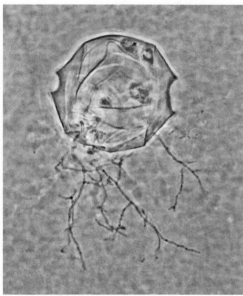

FIGURES 5.4 AND 5.5. Once the zoosporangium matures, the baby zoospores burst forth, leaving an empty husk behind. Photographs and microscopy by Joyce Longcore.

quickly depart from the narrow field of the microscope's vision, I became acutely aware of my own myopia. Veering off at wild angles, each zoospore departed from its parent zoosporangium on its own line of flight. These desiring machines constantly search for suitable places to encyst and grow, places where they can anchor, become new zoosporangia, and start the cycle again. These young chytrids are relentlessly probing the cosmos beyond their idiosyncratic umwelten. Looking for new sites to create spheres, chytrids are ever on the lookout for new beings and things to grasp with their rhizoids, seeking to enfold novel entities into emergent phenomenological bubbles. Once lodged onto a stratum, after they find an advantageous place on it, they experiment with the opportunities it offers.[12]

Joyce keeps chytrids isolated in sterile flasks so that she can describe stable morphologies in taxonomic papers. Selecting out pure strains, keeping each in a sealed tube, she has created isolates—genetically homogenous populations of living fungi. When chytrids are surrounded by other entities, when they live in worlds with other beings and things, they often inhabit different ontologies. Novel structures emerge, which are never seen in the isolated cultures, amid multispecies intra-actions.[13] In contrast to common assumptions about "interactions," which presume the prior existence of stable and independent entities, Karen Barad's notion of intra-action represents a profound conceptual shift. Multispecies intra-actions transform organisms, and even create their modes of existence, amid their encounters with others in common worlds. "It is through specific intra-actions," writes Barad, "that a differential sense of being is enacted in the ongoing ebb and flow of agency."[14] In other words, as chytrids transform themselves with intra-active entanglements, they become ontological amphibians. While Sloterdijk imagines that all nonhumans are trapped in bubbles, bounded worlds of perception and action, it is clear that chytrids routinely break through to ontological cagelessness. Arguably, these sudden breaks make chytrids more amphibious than humans.

The new chytrid species that Joyce Longcore described in 1999, *Batrachochytrium dendrobatidis*, has become infamous for destroying the worlds of literal amphibians. This microbial pathogen is eliminating entire species of frogs, salamanders, and caecilians in diverse corners of the planet. It infects the skin of amphibians, a porous organ they use to "drink" water and absorb important salts (electrolytes). With intense infections, the skin stops functioning properly. Low electrolyte levels cause the heart to stop beating and the death of the animal.[15] In pure culture, this chytrid

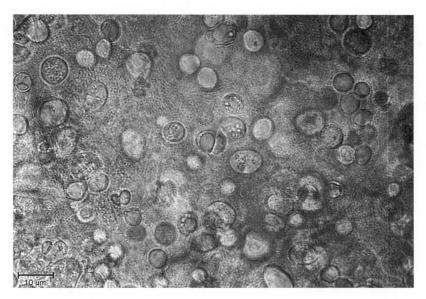

FIGURE 5.6. Zoosporangia of *Batrachochytrium dendrobatidis* embedded within frog skin. Photograph and microscopy by Joyce Longcore.

takes on a stereotypical form—dark spheres within clear bubbles. When infecting frogs, it takes up residence within individual frog skin cells. Electron micrographs of the frog skin reveal that it becomes a rough and bumpy landscape pockmarked with round holes.

Tiny lesions in the skin of dead frogs were independently discovered by Karen Lipps, then a PhD student working in Central America, and Lee Berger, who was also working on her PhD in Australia. Transmission electron micrographs were made of skin from poison dart frogs kept in the National Zoo in Washington, DC. Don Nichols, a veterinary pathologist at the zoo, suspected that they were microscopic fungi, and the images worked their way through the mycological community until they landed in the Longcore lab. "They got lucky with their electron micrograph," Joyce told me. "One shot happened to be right through a zoospore and it was easy for me to see taxonomic clues. When I saw the images, I knew right away that they pictured a chytrid species." When the *New York Times* printed another microscopic picture of this chytrid infecting amphibian skin, from a different frog population, she could see that it was the same thing. Everyone began to speculate that this yet-unidentified chytrid was running wild in a global pandemic.

The paper in which Joyce and colleagues formally named *Batrachochy-*

trium dendrobatidis as a new species had been cited over seven hundred times as of July 2015. Amphibian biologists who work to be precise with their language, those who avoid calling it "the chytrid fungus," usually refer to the disease by its initials, *Bd*, since *Batrachochytrium* is too difficult to say. In the past few years, researchers began to realize that much genetic heterogeneity is wriggling within this seemingly stable species identification. Nonvirulent strains of *Bd* have been discovered in populations of frogs from remote places in Asia.[16] A genetically heterogeneous global strain, now known as the deadly "pandemic lineage," has also been described.[17]

Joyce has assembled the largest collection of *Bd* isolates in the world. Rather than maintaining living cultures, she keeps most of her *Bd* frozen in liquid nitrogen. This is not for reasons of biosecurity—she is not worried about these frog pathogens escaping into the environment. She keeps living cultures of her other chytrid isolates in refrigerators, which already have enough biosecurity, she says. The *Bd* is kept frozen to ensure its genetic stability. If anyone needs a living *Bd* isolate, she can simply defrost the samples and reanimate the cultures.

As a taxonomist, as someone whose core work involves ordering organisms, Joyce is remarkably articulate about the contingent nature of categories. "Species are human constructs," Joyce said. "They aren't things unto themselves. For nonmycologists, for people who work on animals or plants, a species is a group of interbreeding organisms. But there are a lot of organisms in the world that don't have sex." Joyce is well aware of how human concerns shape classification projects. Economic interests and political concerns are constantly torquing existing categories and bringing new ones into being.[18] When there is active interest in a given form of life, categories proliferate. "Pathogens require different descriptors," Joyce told me. They need more specific names. It all depends on human need and use."[19]

Joyce keeps a diverse collection of undescribed chytrids isolated in sealed tubes with secure screw lids. Each Joyce E. Longcore (JEL) isolate receives a number. For example, JEL 352, a chytrid with a "ghostly veil" of wall material, was found in detritus collected from the rain forest canopy along Butcher's Creek near Malanda, in northern Australia. JEL 569 was collected from a pile of horse manure in Maine. With over a thousand chytrids that could be described as distinct species, Joyce and the handful of other specialists need to be strategic about where they focus their attention and scarce resources. "Some chytrids simply don't need to be named," Joyce said. However, even in the absence of pressing agricultural, ecological, or medical concerns, some chytrids earn a name simply for novelty's sake. In 2009, Rabern Simmons, then Longcore's PhD student, described JEL 569, the isolate that was collected from a mound of horse manure.[20] He called it *Fimicolochytrium jonesii*. This Latin name, Rabern explained, means "the horse-shit chytrid of Jones," in honor of a certain Kevin G. Jones—a botanist, mycologist, and associate professor of biology at University of Virginia's College at Wise.

Bd is classified as a member of the Rhizophydiales, a diverse taxonomic order of fungal parasites and decomposers that live in fresh water, marine environments, and soil. Biologists are studying the life history of *Bd*'s close kin to better understand how this disease has come to infect frogs. Rhizophydiales have never been observed having sex in pure culture. But the data from genomic sequencing of different *Bd* isolates suggest that the genetic components of this species are fluid, ever changing. Gene suites, and possibly entire chromosomes, have been lost and gained in different *Bd* strains.[21] Chytrids are clones. Each zoospore is a copy of its parent zoosporangium, usually with only slight genetic mutation. Rather than having straight-up sex, Joyce suspects that *Bd* clones have had a number of "parasexual" events in the recent past. Parasexuality in fungi is messy. According to a definitive account, published in the *Annual Review of Phytopathology* in 1969, parasexuality involves "genetic recombination in which there is no fine coordination between recombination, segregation, and reduction."[22] Some chytrids in other orders do have observable sex. This happens when two zoospores encyst next to one another and send out rhizoids that touch one another. If compatible, the rhizoids fuse. The cell nuclei, which contain the organism's DNA, intermingle as the fused rhizoid begins to swell. This mass then turns into a resting spore—with a thick cell wall resistant to temperature extremes and drying—with a novel mixture of genetic resources. Some species have genders, usually indicated by +/– in the literature, while others are gender neutral.

Sexual or parasexual encounters among *Bd* zoospores have never been seen. But sometimes zoospores develop in laboratory cultures with two distinct nuclei. Sibling spores, from the same zoosporangium, will sometimes fail to divide and thus have twice the genetic material as normal. Longcore's former graduate student, Rabern, has named these oversized spores "Lennies," after Lennie Small—a man of large stature but limited mental abilities—in Steinbeck's *Of Mice and Men*. Venturing into the realm of pure speculation, one might suppose that these lumbering animalcules could be the source of the genetic variability and ontological indeterminacy in *Bd* fungi. Incestuous parasexual encounters between siblings, or even other chytrids from distantly related strains, might occasionally germinate hopeful monsters—new generations of chytrids with a novel suite of genetic resources.[23]

Genetic sequencing suggests that *Bd* might not be a recent emergent disease, but that it became a distinct species approximately 24,600 years ago.[24] Evidence of much genetic variation in *Bd* has also been reported in studies of its genome: some strains have just one copy of each chromosome, while others have two, three, or even four copies of gene suites. Chytrids are thus microbial clones with a shifting sense of self. To paraphrase Brian Rotman, these microbes are sometimes beside themselves with glee and dissolution, intermittently present to themselves, becoming multiple and parallel, each of them a paraself.[25] Paraselves have an auxiliary relationship to the self. These subordinate selves are often irregular, disordered, or improper. Paraselves are like the figural parasites described by Michel Serres—they are jokers, wild cards, taking on different values depending on their positions.[26] Swarms of disorderly chytrid paraselves have become novel parasites, finding exploits that have opened up entry into new hosts and new worlds.

Bubbles of *Bd* are frozen in the Longcore lab to limit promiscuous liaisons that might result in such unexpected genetic becomings. Maintenance of these frozen bubbles is dependent on an industrial supply chain—the periodic refreshing of liquid nitrogen, which comes on a delivery truck every week. In addition to *Bd*, hundreds of other chytrid species are kept as isolates in the refrigerators, incubators, and liquid nitrogen cryobanks of the Longcore lab. Each isolate tube has a number that maps onto handwritten entries, with details about its origin, kept in dog-eared notebooks. In contrast to herbarium samples neatly mounted leaves on paper sheets, or zoological collections with animals preserved in jars of formalin, this collection of living chytrids depends on Joyce and her students to keep them animated. Without an institutional commitment to maintain a complex system of life support technologies, without an endowment to ensure continued maintenance routines, the future of these animalcules is shot through with uncertainty. The bubbles being created in the Longcore lab, where spheres are proliferating inside spheres, are fragile and needy of care, just like the Amphibian Ark. "Spheres are constantly disquieted by their inevitable instability," asserts Sloterdijk. "Like happiness and glass, they bear the risks native to everything that shatters easily."[27]

Chytrids are difficult to maintain in the laboratory. When mixed with all sorts of other critters, when living in gross culture, they are often overrun by bacteria and filamentous fungi. The living collection in the Longcore lab is dependent on constant renewal—colonies must be regularly transferred into new tubes, with fresh media. Most isolates need to be transferred every one hundred days. Others can be left to their own devices longer. In August 2012, I helped Joyce create bubbles for future generations of chytrids. Care for her chytrids takes place in the wet lab, down the hall from the office where her microscopes are housed and where most of her isolates live in a refrigerator. Taking tubes with the established isolates out of the refrigerator, the parents, I carried them down the hall in a small rack. Droplets of water quickly began to condense on the tubes. Another rack with new tubes, ready to be seeded with zoospores, was waiting at room temperature.

Following Joyce's protocol, I began creating bubbles of sterility, purified spaces nested inside one another. I shut the wet lab window down to a crack, minimizing the flow of air, fungal spores, and floating bacteria into the room. Flicking a switch, I turned on a blower, a strong fan inside a sterile hood, to establish an air curtain separating the rest of the room from a small confined space where I would be transferring the chytrids. Firing up a Bunsen burner, I sterilized a metal wand—a long thin piece of wire bent in a right angle at the end—by holding it in the flame until the metal glowed orange. A red plastic rack held the tubes that would receive the spores. Each of these empty tubes had an isolate number written on the side and was already filled with the most appropriate agar medium. Taking my first tube off the rack, I unscrewed the lid halfway and propped it up sideways on a Sharpie marker, a makeshift perch, resting inside the hood. Next I poured a small puddle of alcohol out on the floor of the hood and scrubbed it down with a chem wipe towelette.

Pulling out the appropriate parent tube from a black plastic rack, and wiping off the layer of condensation with my hand, I unscrewed the cap and placed it upside down on the freshly sterilized surface. I kept the tubes sideways when open, to minimize the chance that aerial spores would get inside. The lids of the tubes, the threads where the caps screw down, are "sacrosanct," in Joyce's words. If accidentally touched, the threads should be held to the flame of the Bunsen burner. Some spheres are porous, writes Sloterdijk. They retain "a certain willing openness to seduction by outside elements."[28] Other bubbles, like those cocreated with chytrid isolates, depend on performing the techniques and tactics of immunology. Strict policing of boundaries is necessary, with high stakes and arbitrary distinctions, to keep out suspected enemies while catering to the needs of allies.[29]

Reaching inside the tube with the metal wand, I cut out a small square of agar dotted with visible chytrid colonies from the open tube. Impaling this square with the wand, I teased it out of the parent tube and placed it—spore side down—inside the new tube. After screwing on the cap, returning it to the rack, and flaming the wand in the Bunsen burner, I took out the next tube and began the process again. A week after my visit to the lab, Joyce Longcore wrote me an e-mail: "This morning I checked the cultures that you transferred and found only two tubes with contamination—not bad for a first-timer! We will continue with the culture transfer today."

Sloterdijk suggests that "spheres are the original product of human coexistence—something of which no theory of work has ever taken notice." Outlining the contours of his theory of human exceptionalism, he claims, "These atmospheric-symbolic places for humans are dependent on constant renewal."[30] Imagine, for a moment, a world without chytrids and their microbial kin—a world with growing piles of horse shit, mountains of shrimp exoskeletons, with onion skins and pollen grains that will not decay. Certainly no theory of work has ever taken notice of chytrid labor.[31] Our own bubbles of comfort and happiness are dependent on chytrids, and a multitude of other microbes, for constant renewal. Sloterdijk's sphere, a lonely umwelt where humans are in love only with themselves, must be punctured to trouble the immemorial security of his self-blown bubbles of illusions.[32] Chytrids shape our own ontologies. As they create phenomenological bubbles, disrupting established boundaries with the penetrating grasp of their rhizoids, new worlds emerge around them. Chytrids constrain codependent critters in webs of interdependence. They choreograph moves of other agents in an intra-active dance.[33] Coproducing worlds, they form and transform ontologies in concert with others.[34]

Batrachochytrium dendrobatidis, the deadly chytrid commonly known by its initials, *Bd*, is already troubling the security of many animals that are loved by people. Shortly after it was described as a species by Joyce Longcore and colleagues in 1999, a search was mounted for its origins. The pattern of its spread initially seemed clear: following the emergence of pathogenic strains of this microscopic fungus, it moved like a slow wave around the planet, leaving a trail of extinct and endangered amphibians in its wake. Some began to speculate that the spread of *Bd* was linked to *Xenopus*, the African clawed frog, a model organism that has long been used as a test subject in laboratories around the world. Since *Xenopus* can harbor this fungus with no readily perceptible ill effects, people began linking this adaptable frog to the mass extinction of other amphibians. Following *Xenopus* through historical archives, I began studying how this cosmopolitan animal was enfolded into biotechnical worlds in the early twentieth century. Partnering with artists and scientists, I conducted a "performative experiment"—interrogating the hypothesis that *Xenopus* played a role in an out-of-Africa story of disease emergence.

CHAPTER SIX

XENOECOLOGIES

. . .

While many frogs have become endangered in an era of extinction, some amphibians are flourishing in emergent worlds. *Xenopus laevis* is one such frog. It refuses to distinguish between the domains of nature and culture. In southern Africa, the original home of *Xenopus*, they can be found in all kinds of water bodies: from rivers and lakes to dammed reservoirs, from swamps to flooded pits, ditches, and wells. They prefer nutrient-enriched stagnant water but can also be found in fast-flowing streams and fish farms. During periods of drought, *Xenopus* burrow deep into the mud and enter a state of suspended animation. Amid rain storms and flooding, thousands of these frogs have been observed migrating to new wetlands. "It is extraordinarily resistant to disease and infection," writes J. B. Gurdon of the Wellcome Trust in Cambridge, and is associated with "an exceptionally rich fauna of parasitic organisms." Among other diverse parasites, this robust frog can also host *Bd*, the pathogenic chytrid, without becoming noticeably sick.[1]

The food of these omnivores "appears to include everything available in the aquatic environment," according to a definitive review by R. C. Tinsley and colleagues. Aquatic insects, crayfish, minnows, worms, and other macroinvertebrates are among their staples. Fish, birds, and other amphibians have also been recovered from the guts of *Xenopus* in the wild. The name of this creature means "strange foot" (*xeno* = strange, *pus* = foot) in Latin, while commonly it is known as the African clawed frog. Claws on its powerful hind legs are used to rake and tear at food. "Groups of *Xenopus* may attack the same prey and can tear the body into fragments," continue Tinsley and colleagues. The frogs use their strong hindlimbs and sharp claws to rake carcasses—sometimes large mammals like dead horses, cows, or elephants—and then fork the meat

FIGURE 6.1. An African clawed frog (*Xenopus laevis*) resting underwater in an aquarium at Ueno Zoo, Tokyo. Photo by Peter Galaxy.

into their mouths with their small forelimbs. "The offspring of *Xenopus* may make a significant contribution to the nutrition of the parental population. . . . Cannibalism enables the adult population to exploit a nutrient resource which could not be utilized directly." The tadpoles, which eat algae and other microorganisms, can become food for adults—thereby allowing the frogs to survive in bodies of water that are otherwise devoid of food.[2]

Despite these feeding habits that trouble human norms, pet owners who keep the frogs in their homes report that they are quite gentle. They have no teeth. "I like to dangle my finger in the water," one *Xenopus* aficionado told me. "Loretta loves to lunge up from the bottom and gum at my finger with her mouth." Part of what has made this frog a popular pet is its ability to survive extreme neglect. Adult *Xenopus* can live without food for extended periods of time. This "remarkably robust constitution" also helped make *Xenopus* into a common laboratory animal. Researchers have monitored physiological effects of starvation for up to twelve months, and hypothesize that the frogs can stay alive much longer.[3] "It is a little difficult to say authoritatively how often they should be fed," wrote

Edward R. Elkan in 1938. "They will certainly accept food every day, but they never show any signs of being particularly hungry, and if not fed for a week they do not seem to be the worse for it."[4]

THE FROG PREGNANCY TEST

Xenopus captured the imagination of Elkan, and many other researchers of his day, shortly after scientists in South Africa discovered it can perform a surprising trick: this amphibian can be used to divine pregnancy in humans. The methods for conducting the *Xenopus* pregnancy test are simple. They were described in plain language by Elkan, who popularized the test with an article in the *British Medical Journal* in 1938: "The test proper starts with the collection of the urine. It seems wise to limit the patient's intake of fluids so far as possible. . . . Some 6 ounces of morning urine are collected in a clean—not necessarily sterile—bottle and sent to the laboratory." The article goes on to describe different parts of the frog's anatomy where a small amount of urine can be injected: one popular injection site was in the thigh of the hind leg (in the dorsal lymph node), while others preferred the frog's belly (peritoneal cavity). "The frogs are slippery and difficult to hold," notes Elkan. "I find that the easiest way of dealing with them is to hold them in a coarse meshed net and to inject through the mesh." The urine of pregnant women contains a hormone called human chorionic gonadotropin (HCG). When HCG is injected into a female *Xenopus* frog, she will release eggs within six to twenty-four hours.[5]

Following *Xenopus* through historical archives and to contemporary scientific laboratories reveals that this frog helped change what it means to be human. In addition to enabling people to reliably detect pregnancy at an early stage, these frogs were tied to major shifts within the realms of kinship, psychology, ethics, politics, and economic enterprises. New methods for using and exploiting the biotic properties of *Xenopus* frogs also changed its mode of being in the world. The body of this frog species, and its conditions of life, were torqued by human dreams and emergent biotechnology schemes. In the 1930s and 1940s, citizens worked with scientists to popularize the *Xenopus* pregnancy test. Reproductive rights activists with Planned Parenthood joined researchers to advocate for adopting the frog pregnancy test as a clinical standard. "Some of the objections to this test raised by earlier observers seem to have become invalid," Elkan writes. "The early detection of pregnancy remains important from the psychological and the gynecological points of view. A method which allows the diagnosis to be made within a few hours should be welcome."

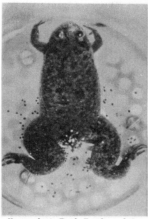

Xenopus laevis Daud, female. Note valves on both sides of external opening of cloaca.

Xenopus laevis Daud. Female ovulating after injection with pregnancy urine.

Xenopus laevis Daud, male. Note absence of valves at external opening of cloaca.

Test jar for xenopus pregnancy test. The center of lid is perforated. The toad sits on a platform: the eggs fall to bottom of jar. Note typical attitude, holding nostrils above surface water.

FIGURE 6.2. "The Xenopus Pregnancy Test," reprinted from E. R. Elkan's 1938 article.

Scientific work in the realm of endocrinology took place alongside a popular mass movement aimed at "challenging the traditional limits of people's control over their own lives."[6] As women were liberated, as they embraced emergent birth control technologies, diverse animals were kept in cages. Alongside frogs, mice and rabbits were also used for pregnancy testing. In the late 1920s, a German chemist, Selmar Aschheim, helped design the "A-Z test," which involved repeatedly injecting five female mice with a woman's urine over several days. The mice were killed, dissected, and then examined to see if their ovaries were enlarged. Swollen ovaries signaled a pregnancy.[7] The Friedman test substituted rabbits for the mice in the late 1920s and led to the association in popular culture of rabbit killing with pregnancy testing. With the rabbit test, human urine was injected into an unmated female bunny and the animal's ovaries were later checked for morphological changes.[8] The rabbit died no matter the result. The *Xenopus* pregnancy test became a more ethical and economical test when compared to the rabbit or mouse test. The frogs were not killed and, if treated well, lived up to fifteen years in labs. They could be reused for pregnancy testing every couple of months. But, since some lab technicians were uncomfortable handling frogs, many continued to use the rabbit test.

Correspondence among key figures who helped popularize the use of frogs over other pregnancy tests foreshadows later concerns by conservation biologists. Margaret Sanger, one of the key visionaries who founded Planned Parenthood, was instrumental in funding early research involving the *Xenopus* pregnancy test. In a personal letter to Margaret Sanger, dated 1951, Edward R. Elkan wrote:

> I should think that where bullfrogs can be raised in such large numbers in a lake, suitable accommodation should also exist for *Xenopus*, a toad [*sic*] which is by no means very demanding in its claims for a suitable environment. The best thing would be if someone who owns a suitable estate, could be induced to free a lake or a swamp from major natural enemies (large fish like pike, carp etc.) and then to release there some pairs of *Xenopus*. If these can maintain themselves in the difficult climate of the African continent I see no reason why they should not have an even better chance in the ideal climate of California.[9]

Decades later, the conversation shifted from promoting the life of this useful frog to considering the ecological ramifications of its worldwide spread. In 2004 Ché Weldon, a South African scholar, published the first peer-reviewed article linking the *Xenopus* pregnancy test with the spread of *Bd*, the pathogenic chytrid fungus. Weldon proposed the "Out of Africa" hypothesis to account for the spread of this disease. Taking tissue samples from amphibian specimens in South African museum collections going back to 1871, he found that the earliest frog testing positive for *Bd* was collected from the western Cape in 1938, just as the pregnancy test was becoming popular. Trudging through swamps and waterways of South Africa, Weldon also tested contemporary populations of *Xenopus* for the parasitic fungus. "The incidence in wild frogs was highly variable, [from] being absent in some of the populations in the Western Cape to being present in 89.7% of animals from the Northern Cape," Weldon reported in his PhD dissertation. "The average prevalence for [wild frogs] is 28.6%." Going through archival records from businesses collecting frogs in more recent years (1998–2004), Weldon also found that over ten thousand *Xenopus* frogs were exported annually from South Africa for scientific uses in over thirty countries. According to figures from the U.S. Fish and Wildlife Department, 1,262,468 *Xenopus laevis* frogs were legally imported to the United States from 2001 to 2009.[10] By transporting *Xenopus* around the planet, Weldon surmised, humans have inadvertently helped spread a plague of frogs.[11]

A PERFORMATIVE EXPERIMENT

Outbreak narratives often follow a predictable script.[12] Africa, long imagined as "the diseased continent" in popular culture, also often appears in the epidemiological literature as the source for diverse maladies—like the Ebola virus and HIV/AIDS.[13] In July 2012, at a time when journalists were linking the outbreak of chytrid to the *Xenopus* pregnancy test, I set out to test Weldon's Out of Africa hypothesis with what Dehlia Hannah calls a "performative experiment." Assembling an interdisciplinary team—consisting of an expert in frog husbandry (Mike Khadavi), a new media artist (Grayson Earle), two ethnographers (Charlie Nichols and Lisa Jean Moore), and a feminist philosopher (Dehlia Hannah herself)—we offered free pregnancy tests with live frogs. Using DNA test kits from the Amphibian Disease Laboratory at the San Diego Zoo, we also offered free fungus tests for the *Xenopus* frogs living in New York City. We intended to see if these frogs were infected with pathogenic chytrids.

New York City pet stores were selling *Xenopus* cheaply in the summer of 2012, with the common names of underwater frogs or African clawed frogs. We invited members of the public to purchase their own *Xenopus* and learn how to perform the pregnancy test in their own homes. By reenacting the *Xenopus* pregnancy test and crowd-sourcing the collection of frogs, we intended to catalyze conversation about elusive ecological entanglements and obscure technical procedures. The free pregnancy tests with live *Xenopus* frogs were to take place in the Multispecies Salon at the Proteus Gowanus gallery in New York City, where the empty *Utopia for the Golden Frog* was on display. We began telling convoluted tales about multiple species of ontological amphibians that had reportedly driven literal amphibians extinct.[14]

Dehlia Hannah coined the term "performative experiment" to characterize art interventions at the edge of standard scientific practices. "Distorting and displacing conventional technical and scientific practices," writes Hannah in our coauthored essay, "our aesthetic interventions were calculated to illuminate features of experimental design, materials and methods, and conventions of interpretation that tend to escape attention in the practice of everyday science."[15] Posting a free classified ad on *Brokelyn*, a web magazine for New York City residents, we offered to demonstrate basic laboratory skills that would enable ordinary people to conduct pregnancy tests in their own homes:

> South African researchers made a surprising discovery in 1934. When they injected the urine of a pregnant woman into a frog it laid hun-

dreds of eggs. This trick earned members of this frog species, *Xenopus laevis*, a ticket around the world. The frog pregnancy test quickly became a standard test used by medical doctors as well as organizations like Planned Parenthood. Free pregnancy tests will be performed in the Proteus Gowanus gallery on Friday, July 6th. We will also share some simple laboratory skills. Frogs will be given away free to good homes with people who are ready to use knowledge of this Do-It-Yourself test. No permanent harm is done to the frogs in testing and they can be reused multiple times.[16]

Journalists discovered our classified ad and began to report on the upcoming gallery happenings. "Think you might be pregnant?" asked the *Huffington Post*. "All you'll have to do is inject your urine into a frog. Cool?"[17] *Metro*, a free daily newspaper with over 700,000 readers, published a story about the test showcasing ecological issues: "The frog pregnancy tests are part of a larger exhibit that aims to shed light on how the tradition may have contributed to the demise of frog populations as a result of a fungus spread by the tests. The African clawed frog is not affected by the Chytrid fungus, but is a carrier of it and often spreads it to other frog species."[18] *Scientific American* asked its readers, "Is your store-bought frog carrying a deadly secret?"[19] These media accounts catalyzed conversation about how *Xenopus* frogs and *Bd* fungal spores might be exploiting faults and fissures within emergent ecosystems and established biotechnical assemblages. The exhibit at Proteus Gowanus showcased nascent multispecies associations and illustrated how amphibious critters were supplanting deeply rooted ecological communities.

After reading newspaper accounts about our upcoming exhibit, activists with People for the Ethical Treatment of Animals (PETA) became outraged about our intention to conduct pregnancy tests with live frogs. Misconstruing the protocol of the test and some of the issues at stake, a PETA spokesperson told reporters and supporters, "The frogs would be injected as many as 60 times each, resulting in significant pain and distress."[20] Days before our scheduled opening at Proteus Gowanus, this anonymous PETA spokesperson told a journalist at *Metro* newspaper that we planned to intentionally infect frogs "with a deadly fungus for the sake of so-called 'art'" and that we were "potentially risking the lives of every frog in the area" by doing so.[21]

This nameless PETA representative had the details of our intervention mixed up. Rather than infecting frogs with a deadly fungus, we would use sterile swabs from the Amphibian Laboratory of the San Diego Zoo to

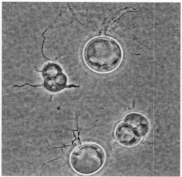

FIGURES 6.3 AND 6.4. *The Xenopus Pregnancy Test* was part of a group exhibit called *Future Migrations*, curated by Krista Dragomer, which probed how "movements are affecting our future on the planet, bringing crisis and calamity aplenty." We displayed living *Xenopus* frogs in mason jars alongside pictures of chytrid zoospores from Joyce Longcore's lab. Photographs by Rashin Fahandej and Joyce Longcore.

see if any of the frogs in our care were asymptomatic carriers of *Bd*, the pathogenic chytrid. Following the confusion generated by PETA's statements, we simplified our intervention at Proteus Gowanus. We canceled the pregnancy tests in the art gallery and instead offered free chytrid tests to anyone who wanted to keep *Xenopus* frogs as pets. Members of the public brought us frogs purchased from a diverse array of pet stores in New York City—from Petland Discounts in Bedford-Stuyvesant, Bob's Tropical Pet Center in Ridgewood, Brown Stone Aquarium in Carroll Gardens, and Petqua in Manhattan. We also purchased a couple of frogs from Xenopus Express, a specialist supplier that provides laboratories around the country with experimental animals. These frogs, which were guaranteed to have "high quantities of firm oocytes," cost $26 each plus $50 in shipping with UPS by Next Day Air delivery from Florida.[22]

Our *Bd* test kits from the San Diego Zoo contained fine-tipped swabs, which looked like long Q-tips. Swabbing frogs involves a relatively simple procedure. After donning gloves, to protect the frogs from soaps and lotions commonly on human hands, one must endeavor to catch the animal. A small net, available at pet shops that sell fish, can be helpful. Once the frog has been secured (I found it easiest to hold it in my left hand while swabbing with my right), the protocol commences. "Using a single swab, gently swab the ventral surfaces of the skin approximately 20–30 times. Target areas to include the pelvic patch (5 passes with the swab), ventral thighs (5 passes each side with the swab) and toe webbing (5 passes on each foot)." After rubbing twenty *Xenopus* frogs in the prescribed pattern, we bottled up each long Q-tip in a small plastic vial and FedExed the package to the San Diego Zoo.

While waiting for the results, my team of collaborators started making house calls in Brooklyn—making good on our offer to perform frog pregnancy tests, free of charge. We followed the instructions in Edward R. Elkan's 1938 article: collecting morning urine in a clean—but not necessarily sterile—container, and then injecting a small amount into the dorsal lymph node on the thigh of the frog's hind leg. Each frog used in pregnancy testing was injected only once, not sixty times as PETA alleged. With a nod to Donna Haraway, who has written about the possibility of "sharing suffering" with laboratory animals, we also injected ourselves with saline solution. The injections stung a little. They felt like a tuberculosis test, in which a little bit of liquid is injected under the skin. But we had no way of sharing the alarming experience of being captured and held by a giant, either for the benefit of humans who wish to detect pregnancy or for a procedure, like fungal swabbing, that has the interest of the tested organism in mind.

Sharing suffering, for Haraway, involves "the practical and moral obligation to mitigate suffering among mortals—and not just human mortals." Rather than engaging in "the heroics of self-experimentation," which is a common trope in the history of medicine, sharing suffering entails attention to "the most vulnerable lab actors."[23] Unlike harboring a fantasy of ending all suffering by laboratory animals, or even claiming to "feel the pain" of individual frogs in our care, our performative experiment aimed to expose standard laboratory methods involving experimental animals. Following the outcry from PETA, I began to dig deeper into the biological literature, to explore other uses and abuses of this model organism.

Once viewed as "primitive" creatures by experimental biologists, *Xenopus* frogs were thought to be unable to experience feelings of pain or fear. As a result, they were subjected to invasive experiments that likely would have been blocked by university ethics committees had they been proposed in rats, monkeys, or other more "evolutionarily advanced" creatures. Early researchers who studied *Xenopus* frogs in the lab assumed that they inhabited a limited umwelt—that they lived in an impoverished world, an ontological cage. Indeed, the frog has a limited repertoire of behaviors that can be elicited in response to specific stimuli. One researcher told me how he switched from studying bird song at the beginning of his career to working with this frog because it is "easier to take the whole system apart and figure out how it works." He told me about growing an eviscerated brain in a petri dish to study *Xenopus* mating calls. "You take the brain out of the frog . . . and the brain will sing in the dish. You put an electron on the nerves that would normally go to the vocal organ and the brain is happy, singing in the dish."

According to a laboratory manual published in 2010 by the Taylor and Francis Group, "*Xenopus* have all of the neuroanatomical pain pathways as seen in mammalian species, and thus, like mammals, they are capable of experiencing pain. . . . Heightened awareness for the welfare of earlier-evolved laboratory species has prompted increasing inquiries by institutional animal care committees, investigators, and veterinarians."[24] Research indicates that adult *Xenopus* frogs can also retain long-term memories and are "capable of learning rules in their environment." Adult frogs are able to act on memories acquired when they were tadpoles.[25]

While *Xenopus* have poor eyesight, and a short list of sounds they use for vocal communication, they are able to perceive and act upon other dimensions of the world to which humans are completely blind. To our eyes, adult *Xenopus* appear to have a complex pattern of stitching around the midsection of the body (see figure 6.1). These stitches are actually part of a sensory organ, the lateral line, which detects movement and vibration patterns in water. *Xenopus* frogs thus intra-act with vibrant matter. Their agency emerges with ad hoc configurations and reconfigurations of ecological, corporeal, and molecular forces.[26]

As experimental animals, frogs continue to suffer, serve as scapegoats, and die so that we might live. Like OncoMouse, the transgenic animal created by Du Pont as a model for studying breast cancer, *Xenopus* has become "a kind of machine tool for manufacturing other knowledge-building instruments in technoscience."[27] The many talents of this frog—the clever tricks of the sensory system, the hormones, and the genes of this animal—have transformed it into what Donna Haraway calls a companion species, a critter that makes "life for humans what it is, and vice versa."[28] These strange creatures have helped humans craft new social realities and science fictions.[29] They have helped generate new modes of being human that are dependent on complex entanglements with animals, ecosystems, and emergent biotechnologies.[30]

ONTOLOGICAL CHOREOGRAPHY

By reenacting the frog pregnancy test, we were trying to learn more about how *Xenopus* has taken advantage of exploits, openings into new technological and environmental worlds.[31] We were also trying to learn something about our bodies, and ourselves. We were interested in how these living beings have helped humans extend our phenomenological and ontological bubbles.[32] "Our performative experiment interrogated how the phenomenon of pregnancy itself involves intra-active ontological choreography with members of multiple species, as well as social and material technologies," in the words of Dehlia Hannah and our other collaborators.[33] Ontological choreography, as originally applied to reproductive technologies by Charis Thompson, involves "the coordinated action of many ontologically heterogeneous actors in the service of a long-range self."[34] Working at assisted reproductive technology clinics in the 1990s, Thompson described "a deftly balanced coming together of things that are generally considered parts of different ontological orders (part of nature, part of the self, part of society)." She described how different actors were "coordinated in highly staged ways" to produce parents and children.[35]

Our performative experiment exposed the intimate dances of multiple species involved in early reproductive technologies. It also offered a glimpse of the diverse ways that *Xenopus* frogs open up opportunities for us humans to choreograph our ontologies differently. Staging a pregnancy test using a standard, but outdated, method exhibited the historical conditions of possibility under which the pee-stick test became the standard. *Xenopus* frogs and the pee-stick tests both detect human chorionic gonadotropin (HCG) in a woman's urine, a hormone that is released only after an egg has been fertilized by a sperm cell and the blastocyst has implanted in the lining of her uterus. This can happen as early as eight days after conception, and the elevated hormone levels can be detected by a sufficiently sensitive test before a menstrual period is missed. A positive test at a very early stage with a pee stick also often presages an early miscarriage. The *Xenopus* test is slightly less sensitive than the pee-stick test, which is itself less sensitive than a blood test. Therefore the test itself—the chemical technology or animal assay—determines the temporal point in the progression of pregnancy at which a yes or no answer can be given. There is thus an appreciable gap between the biochemical condition of "being pregnant" (with the production of HCG hormone) and the social condition of experiencing oneself and being recognized by others as pregnant—a gap that the technology of the pregnancy test can serve to widen or narrow, depending on how one chooses to choreograph an ontological state.

In the mid-twentieth century, in the decades after the *Xenopus* pregnancy test was first discovered, this promiscuous frog began hopping across disciplinary divides in the biological sciences, enabling choreographers to script diverse new movements. This ontological amphibian transformed economic enterprises, social norms, and ecological communities. After capturing the imagination of researchers studying the hormones involved in pregnancy, *Xenopus* quickly became the model organism for endocrinological studies of hormones. Soon thereafter it became the standard laboratory animal used in cell and developmental biology. "This further popularity," in the words of J. B. Gurdon, "mainly results from the large size of *Xenopus* embryos and cells." The large egg cells (oocytes) of this frog have recently come to be regarded as "living test tubes" for transgenic research. Biologists were delighted to discover that "when the genes of a foreign species, vertebrate or invertebrate, are injected into an oocyte nucleus" they are translated into proteins and assembled into complex structures.[36] The popularity of this research animal continues to grow. *Xenopus* became the first free-living vertebrate species to conceive in an extraterrestrial environment on Space Shuttle Endeavour during a September 12, 1992, mission.[37] Between 1998 and 2009, the number of published studies using *Xenopus* in PubMed, a definitive database of biomedical research, increased fivefold.[38]

While life in laboratories resulted in an uncomfortable existence for millions of frogs, and subjected some to gruesome dismemberment, new opportunities for flourishing also emerged as *Xenopus* played diverse roles in scientific and economic ventures. From South Africa, they traveled through the networks of Planned Parenthood in the United States, to endocrinology and genetics laboratories in diverse corners of the globe, to the pet stores of New York City.[39] These tenacious frogs are currently living in drainage systems along the U.S.-Mexico border, canals in Holland, and underground water cisterns in Welsh castles.[40]

Xenopus frogs are flexible. They are able to inhabit different ontologies in collaboration with diverse choreographers. Despite their poor eyesight and a limited repertoire of physical movements, extra copies of raw genetic material give these amphibians the opportunity to configure and reconfigure themselves at a cellular level. "The animal has enormous genetic capacity, which could contribute to their success," says Darcy Kelley, a biologist at Columbia University whose career has orbited around *Xenopus*. "They have opportunity to use their genetic molecular tools in new ways. . . . Mostly you can't screw around with genes. Because if genes have been around for a long time they are doing whatever they do pretty well. So if you take some gene and screw around with it, it's pretty rare that it does things better. But, if you have spare genes you can screw around with them, what the hell."[41] Many animals, like humans, are diploids, with two copies of each chromosome. *Xenopus* frogs are polyploids, with multiple sets of redundant genes. As tetraploids, octoploids, or dodecaploids, different species of *Xenopus* have four, eight, and twelve copies of each chromosome respectively. With multiple copies of genes required to build a whole organism *Xenopus* can experiment with modes of being.

Signaling systems function in the cells of all organisms, turning genes on and off, forming a subconscious umwelt of sorts. Cell receptors detect an external stimulus. They send chemical signals, activating dormant genes wound up in chromosomes. Genes are unwound, transcribed, and translated into proteins. Proteins are assembled into enzymes. Enzymes catalyze new cascades of material changes and semiotic signals in the cell, in the tissues of the organism, and in the environment beyond. Diverse molecular tools in *Xenopus* may well help frogs grapple with pathogenic chytrids, making this animal a poster child for survival in an era of amphibian mass extinction. In addition to flourishing in contemporary ecosystems, Darcy Kelley speculates that this adaptable frog will outlive the human species: "These animals are incredibly evolutionarily successful," she told me. "You know when the meteorite hit the planet, in Sea of Cortez, and there was this huge extinction event? *Xenopus* survived that just fine. When we are long gone from this planet there will be cockroaches and *Xenopus*."

MULTISPECIES MIGRATIONS

As we produced definitive results about pregnancy among our extended network of friends, and listened to tales about the past, present, and possible futures of adaptable frogs, we waited for other results: the laboratory findings from the chytrid swabs we sent to the San Diego Zoo. As days turned into weeks, people who had adopted *Xenopus* frogs began e-mailing with concerns that their new pets might be sick. One woman wrote, "If my frog has the fungus, what should I do?" We directed her to web pages of amphibian specialists, where baths of diluted antifungal medicines like Lamisil and itraconazole are recommended, but told her to wait. When the results from our swabs were reported by the laboratory, we were met with a surprise: none of the *Xenopus* frogs we tested had detectable *Bd* zoospores. All of our frogs were free of this pathogenic chytrid fungus. Our performative experiment in Proteus Gowanus thus added a new twist to the entangled tale of *Xenopus* frogs and their chytrid companions. Certainly this finding did not directly disprove Ché Weldon's Out of Africa hypothesis. It is difficult to document the presence of an absence.[42] Still, the results of our performative experiment generated new empirical evidence that destabilized the story about an outbreak from "the diseased continent" facilitated by a pregnancy test gone awry. These results offer an opportunity to paint a much more complex picture.

A multispecies story about disease emergence is coming to light with publications in the primary biological literature. A 2012 paper in *PLOS ONE* found that waterfowl can transport chytrid zoospores, reporting that "*B. dendrobatidis* is highly prevalent on geese toes."[43] While the geese show no apparent ill effects, other hosts, like crayfish, became fatally sick. Infections with *Bd* "caused significant crayfish mortality and gill recession" in the electric blue crayfish and the Louisiana mudbug, according to a January 2013 article in the *Proceedings of the National Academy of Sciences*.[44] Researchers surveying frogs in Asia found a very low level of prevalence of *Bd* throughout the continent (2.35 percent infected overall) and concluded that the "infected animals were not clumped as would be expected in epizootic events."[45] In other words, the widespread and low-level infection patterns in Asia suggest that this chytrid has been there for a long time, which would confound the Out of Africa scenario.

Genome sequencing data, from a study published in March 2013, reveal much genetic variability and ontological indeterminacy within different *Bd* strains. One strain of these dynamic microbes, the Global Pandemic Lineage, was highly pathogenic and was killing frogs in Europe, Africa, Latin America, and the United States. Other strains were relatively benign and restricted to particular locales—one was isolated in Brazil, another from southern Africa and Spain; another was found only in Switzerland. The paper states that "it is premature to conclude a geographic location for the origin of Bd."[46]

Even though the frogs we tested in the New York City pet stores were not carrying chytrids, other researchers (in addition to Weldon) have found *Xenopus* frogs with active infections living throughout Africa and in the wilds of California.[47] The American bullfrog (*Rana catesbeiana*), a species exported to international food markets as frog legs, is also an asymptomatic carrier of *Bd*. But neither of these frogs is found in high-elevation areas where most declines and extinctions of frogs have been documented.[48] Some native frogs of California have also helped spread this chytrid, indirectly helping kill other frogs that once coexisted with them in the same environments. The Pacific chorus frog (*Pseudacris regilla*) harbors active *Bd* infections with few side effects and continues to persist in the high-elevation Sierra Nevada where mountain yellow-legged frogs (*Rana muscosa* and *Rana sierrae*) have undergone catastrophic declines. In Panama, in regions where golden frogs once lived, diverse frogs continue to persist even though they are infected with *Bd*: red-eyed tree frogs (*Agalychnis callidryas*), the hourglass tree frog (*Dendropsophus ebraccatus*), Fitzinger's robber frog (*Craugastor fitzingeri*), and the emerald glass frog (*Centrolene prosoblepon*).[49]

Chytrids thus can be something of a gift to some of their amphibian hosts who can offer hospitality with minimal loss. As companions for some frogs, and killers of others, these fungal parasites are transforming established multispecies communities. Together with some cosmopolitan animals, like *Xenopus* and American bullfrogs, and a diversity of native amphibians, chytrid zoospores are shaping emergent ecologies, they are destroying established communities even as they generate new associations. Michel Serres, whose book *The Parasite* celebrates the creative potential of these lively agents, also recognizes that parasitic relationships involve "asymmetry, violence, murder and carnage, arrow and axe." "It might be dangerous not to decide who is the host and who is the guest," concludes Serres, "who gives and who receives, who is the parasite and who is the *table d'hôte*, who has the gift and who has the loss, and where hostility begins within hospitality."[50]

TRAPPING NOMADS

Xenopus has been woven into the ready-made script of invasive species, a familiar plot characterized by Comaroff and Comaroff as "the naturalization of xenophobia."[51] "Killer Meat-Eating Frogs Terrorize San Francisco," screams one headline from Fox News, reporting about the presence of this frog in a pond in Golden Gate Park.[52] "It's like something out of an animal horror movie," according to the *San Francisco Chronicle*. "Killer frogs take over peaceful pond, then after terrorizing and eating everything alive, they start eating each other."[53] Conservationists and environmental activists were taking part in a campaign of killing of their own, trying to exterminate *Xenopus* and other kinds of cosmopolitan frogs that were flourishing in changing ecosystems.[54] Xenophobia led to xenocide.

"Saving nature is a deadly project," in the words of Donna Haraway. As conservationists made high-stakes and potentially arbitrary distinctions between native animal friends and foreign foes, the question was not just, who will live and die?[55] The question also became, how will they die? In San Francisco conservationists proposed a final solution to the *Xenopus* situation. Fearing that these "fiendish amphibians" might "spread their reign of terror across other Bay Area waterways," the California Department of Fish and Game initially proposed to drain the ponds of Golden Gate Park, killing the ecosystem along with the frogs. Instead, park workers began using nets and traps baited with chicken, euthanizing some 2,500 frogs with nerve poison after yanking them out of the pond.[56] While militant environmentalists regarded these nomadic frogs as potentially dangerous and irredeemably destructive, others worked to turn the presence of *Xenopus* into an opportunity. Enterprising businessmen worked to twist death back into new forms of laboratory life.[57] A company called Pacific Biological began adding value to these prolific frogs—they began catching them in the wild and selling them as commodities for use in biological research. Trapping these nomads, the company enfolded the frogs into stable cosmopolitical worlds while facilitating their further cosmopolitan migrations.[58]

Studying the stories of invasive species, and the ecologies emerging around experimental animals, I traced the routes traveled by the actual animals we used in the pregnancy test, the scientific-grade *Xenopus* frogs that were overnighted to the art gallery in New York City. I ended up in Florida, the state where I was born. I wanted to learn more about the movements of these frogs around the United States, to see if they were helping spread chytrid zoospores. The company that sold me the frogs, Xenopus Express, claimed that the facility was periodically screened to make sure that the animals were fungus free. My own independent chytrid tests, on the frogs used for pregnancy tests, confirmed this claim. Still, I wanted to better understand how Xenopus Express was situated within the ecologies of Florida. Burley Lilley, the self-made man who owns and operates Xenopus Express, met me at a local lunch joint in a shopping mall, near his facility about an hour north of Tampa. I began the conversation by asking Burley about how he got into the laboratory animal industry.

"Basically I was the first vendor with a website, xenopus.com," he tells me while eating a hot corned beef sandwich on pumpernickel bread. "I was

FIGURE 6.5. The frogs guaranteed to have "high quantities of firm oocytes" for the pregnancy test were purchased from the 1-800-Xenopus hotline. They were delivered by UPS Next Day Air from Florida to the gallery in plastic Tupperware containers filled with moist packing foam within a cardboard box that had a bright green label: "Live Frogs." Photograph by Eben Kirksey.

lucky enough to get the domain name, and my hotline: 1-800-Xenopus."
With a chuckle, Burley told me about one of his competitors, a former
customer. This dedicated frog enthusiast kept *Xenopus* in his own home
and then tried to sell them to local pet stores. "He and his wife took their
bed out of their bedroom and built tables around the whole room with
fish tanks. Basically, the frogs took over their house. They had to sleep
on an air mattress at night in their living room. As soon as they got up
they would put the air mattress up." Burley told me about other hobbyists who had become close to the frogs in different ways. "They will put
their hand in the water and the frogs will swim right up to their hands,
kind of like a dolphin or a whale. They will eat right out of their hands.
It's the craziest thing."

Burley's facility, which he declined to let me visit, holds hundreds of
thousands of frogs. "For me it is more of a business. I don't really bond
with them," he said. Holding tanks at Xenopus Express have big PVC pipes
that serve as hiding spots for up to twenty or thirty frogs at a time. "They
pile up like cockroaches in those pipes. I wouldn't want to be in there.
'Cause they have to come up for air about every twenty minutes. To be in
the middle of that tube and have to get some air would be kind of creepy.
But they manage to do it in time." Despite having occasional creepy feelings, when he imagines himself living underwater in a frog's body, it is
clear that Burley knows how to care for these curious creatures. "If they
are hanging out at the top of the tank, instead of hiding in the bottom,
you know they aren't feeling well," he said. "Honestly, when they get to
that point, they've either got a couple of things wrong or a couple of diseases. If they are darting to get out of the water, if they aren't happy with
their conditions, then they will try to take off. The same thing happens
when it rains. They go crazy trying to get out of the water."

This moment in the conversation was my opportunity to ask if any
of his frogs had escaped, if there were any wild populations of *Xenopus*
in Florida. Burley said, "I've had two reports of *Xenopus* in Florida. One
was in the Miami River and one was by USF in Tampa. I don't think that
Florida is the optimal climate for them. Because it gets too hot. I know our
frogs and we have to cool their water in the summer. In Britain, France, a
couple places in Europe they are there—but not like big time. I think Sic-
ily it is big time there. But not so much in Florida." Burley's claims check
out with the primary literature. In September 1964, about two hundred
Xenopus laevis were released in the Red Road Canal, in Hialeah, a suburb of
Miami, by a disreputable animal importer who decided to let excess animals loose rather than glut the market, according to the *Quarterly Journal*

of the Florida Academy of Sciences. Despite repeated surveys of this site, they have not been seen since. Nomadic frogs, trying to go elsewhere every time it rains, thus appear to often land in inhospitable terrain. Rather than worrying about frogs getting out, Burley is more concerned about other animals getting in. "Raccoons love frogs. And a lot of the native birds like 'em too, like egrets and the herons."

While many environmentalists in California continue to maintain an a priori distinction between native victims and alien villains, in Florida I found policy makers who were less hasty in passing a judgment of harm. Scarce resources for ecological management meant that the presence of novel animals or plants was being judged amid shifting contingencies, as officials carefully evaluated costs and benefits. Florida is currently home to over 500 species of exotic animals and 1,180 kinds of alien plants, and the state simply does not have the staff to eradicate them all, even if they wanted to. "As many as 40 exotic agricultural pests arrive here each month," according to the Florida Fish and Wildlife Conservation Commission's website. Despite sensational media reports about Burmese pythons and walking catfish, Florida officials remain sanguine about the situation: "Many nonnative species are a benefit to people, such as citrus trees and cattle. Fortunately, of all the exotic species that escape or are released, only a handful will survive and become established. The majority of those few species that survive will probably not have negative effects on native wildlife."[59]

Alongside Burley's business venture, I found permissive laws in Florida that allow ordinary citizens to keep a riotous diversity of organisms as personal pets. As animals escaped or were released, new ecological communities were emerging. Dumpsters at a mobile home park near Daytona Beach were being raided by a multispecies troop of several dozen rhesus monkeys, two pig-tailed macaques, and some squirrel monkeys.[60] Parrots from South America, Indian peacocks, and spiny-tailed iguanas were on the loose in Sarasota. Stories of wild monkeys, rumored to have escaped from the set of the original *Tarzan* films, captured my attention. Following these feral monkeys around the state, I began to spend time with Floridians who have assumed a special responsibility to guard them from political agents who demonize immigrants and exotic animals. While some people simply celebrated the capacity of introduced monkeys to *become wild* in emergent ecosystems, others worked on a more subtle project. Self-appointed monkey protectors helped their primate kin navigate the ambivalent complexity of contemporary political, cultural, and ecological processes.

CHAPTER SEVEN

BECOMING WILD —

. . .

A streetwise rhesus macaque was thrust into the national media's spotlight shortly after it was discovered in 2009, near Tampa, Florida. The monkey was first seen around lunchtime—rooting around in a strip-mall dumpster, picking through the trash from a variety of casual dining establishments: Bennigan's, Panda Express, and Chipotle. "No pet macaques were reported missing around Tampa Bay," reported Jon Mooallem for the *New York Times Magazine*. "There wasn't even anyone licensed to own one in the immediate area." Pet monkeys are usually timid when they escape into unfamiliar urban environments, and often get killed when they race into traffic or grab power lines. This monkey appeared calm, and retreated to a nearby tree for a nap, as people crowded around the dumpster. When a local trapper loaded a tranquilizer dart into his rifle, the monkey "jolted awake, swung out of the canopy and hit the ground running. It made for the neighboring office park, where it catapulted across a roof and reappeared, sitting smugly in another tree, only to vanish again."[1]

Newspaper and television journalists began to track monkey sightings, calling him the Mystery Monkey of Tampa Bay. As some locals began to clandestinely feed him, others expressed outrage—citing public safety and environmental concerns. An enterprising businessman started a Facebook page for the Mystery Monkey, which attracted over 85,000 fans. People began posting photos of their encounters with the monkey, reporting on his brushes with the law, and promoting their own businesses, like the local website www.tampabaynightlife.tv. Supporters cheered the monkey along with their Facebook posts: "Stay safe mystery monkey!" "Keep on the move, don't let them catch you." "Go, monkey, gooooooo!"[2] After several attempts to capture the Mystery Monkey, he apparently learned how to deal with tranquilizer darts. Every time he was darted, the monkey would climb to the top of a tall tree and sleep it off. As the drama continued, *The Colbert Report*, of the TV channel Comedy Central, lampooned the hapless Florida authorities with a special televised feature: "Macaque Attack—1,381 Days of Simian Terror in Tampa."

Following the Mystery Monkey from the national mediascape to suburban landscapes, I helped form an interdisciplinary team to study other rhesus macaques (*Macaca mulatta*) living in Florida's fragmented woodlands. Drawing inspiration from the call for "new forms of collaboration" by the Matsutake Worlds Research Group, we experimented with mixed methods to explore the shared lives of humans and animals who become involved in hybrid communities.[3] Erin Riley and her graduate student, Tiffany Wade, brought the conceptual tools and methodological techniques of ethnoprimatology to the team. If early primatologists presumed that "natural behavior" of our close kin was best studied in "natural environments," ethnoprimatologists focus their research on sites where the social spheres and ecological networks of humans overlap with those of other primates.[4] Elan Abrell, a cultural anthropologist who studies practices of "saving animals," helped situate these wild monkeys within the animal rights movement of the United States. In studying encounters among primates, we asked, *cui bono*, who benefits, when species meet? What is at stake when humans and monkeys become wild together?[5]

Becomings involve new kinds of relations, in the words of Gilles Deleuze and Félix Guattari, that emerge with nonhierarchical alliances, infectious affects, and the mingling of creative agents. Departing from Deleuze and Guattari, whose ideas about "becoming animal" have been critiqued by Donna Haraway for misogyny, fear of aging, and a lack of curiosity about actual animals, we studied the mixed emotions, the contagious delight and fear, at play when primate species meet.[6] As high excitement and high anxiety jumped across the species interface, we considered how the monkeys have flourished amid shifting regimes of power. Mixing the methods of ethology and ethnography, while taking film footage on the fly, we studied the dynamics that emerge when humans and monkeys become wild together.

Australian intellectuals have begun to play with the notion of wildness, one of the keywords of colonialism. Deborah Bird Rose insists that white settlers brought wildness to previously stable places in Australia. She describes some landscapes as "quiet country," in contrast to "wild country" that was created by colonial exploitation or national development. Quiet country still exists for Aboriginal Australians in places where "all the care of generations of people is evident to those who know how to see it." The wild, according to Rose, exists in places trammeled by capitalism, "where the life of the country is falling down into the gullies and washing away with the rains."[7] Our research in Florida took place as a slow wave of home foreclosures swept through the state and as Trayvon Martin, an unarmed African American boy, was murdered by a white vigilante. Wild farm labor practices were exacerbating histories of dispossession in the Caribbean plantation economy, generating precarious modes of life.[8] Florida had become a wild country, where deeply rooted colonial legacies interacted with novel forms of wildness—the "dangerous, risky, and out of control" elements in regimes of overdomestication or hypercultivation described by Sarah Franklin.[9]

Florida's monkeys embody elements of old and new forms of wildness.[10] Aside from humans (genus *Homo*), macaques are the most widespread primate genus—ranging from Japan, to Indonesia, to North Africa.[11] Rhesus macaques have a long history of adapting to new environments. With help from humans, they are becoming wild in new locales. Florida has been described by Laura Ogden as "a densely inhabited and wild area," or an "inner frontier."[12] When people and other primates become wild together in frontier zones, as affects oscillate between elation and fear, situations of unequal vulnerability often emerge.[13] Monkeys become vulnerable when people see them as out of control, while humans who harass wildlife are rarely sanctioned. While some people deliberately provoke dangerous, risky, and out-of-control monkey encounters, others are carefully cultivating a quieter form of wildness. Rather than viewing the monkeys as part of nature that needs to be tamed, supporters understand them as wild animals that need to be guarded and nurtured.[14] These advocates have cultivated relationships with monkeys based on politeness and tact, while remaining open to the possibility that they might escape at any moment.[15]

Wildness is a potent, yet ambivalent term that can be used to make strategic interventions.[16] Theorizing "the wild" is a means of supporting the autonomy and alterity of other species. Wild animals flourish with freedom from controlled (or forced) proximity with humans.[17] Wildness fits, as a concept, with the quickly oscillating affects—involving excitement/fear, elation/trepidation—often associated with encounters among like-minded animals in multispecies worlds. Wildness also fits as grounded theory for understanding contemporary discourse and practice in Florida related to environmental preservation, libertarian politics, and economic systems. Creating a space for quiet forms of wildness, naturalists and citizen scientists are learning how to care for the damaged country of Florida amid long legacies of violence and ecological change. By blocking attempts to cull, kill, and trap wayward forms of life, monkey advocates are opening up new conversations about the politics and categories of belonging.[18]

WILD THINGS

After evading Florida authorities for months, the Mystery Monkey bit a sixty-year-old woman who had been secretly feeding him in Lakewood Estates, a lower-income neighborhood in south St. Petersburg. A few weeks later he was captured following a focused monkey hunt. The bite marks, and witness accounts, suggested that it was a "reactive bite" because the monkey was shocked or scared.[19] Following much debate about his fate, he was given to a privately run zoo in Dade City called Wild Things, where he currently resides. The zoo renamed him Cornelius, a cheeky nod to the ape protagonist in the *Planet of the Apes* franchise. Wild Things tried to take advantage of his celebrity to generate more business, posting messages on his Facebook page, encouraging fans to come visit, commenting on holidays and local events. In response, some of the Mystery Monkey's Facebook friends began donating money to the zoo to build him a bigger enclosure and find him a female companion. Other fans lamented his capture and even openly plotted a jailbreak.

When we visited Wild Things in January 2013, we found that the dangerous and risky elements of wildness had been more or less placed under control. Dade City is a sleepy agricultural town in central Florida, about an hour east of Disney World and other iconic tourist destinations in Orlando. Exhibits at Disney juxtapose the old wild and the new wild: nostalgic visions of colonial frontiers appear alongside fictions about techno-scientific futures.[20] Retro visions of wildness were on display in Dade City. We had difficulty finding the zoo's entrance. After circling through the streets for twenty minutes, we noticed a prefabricated house with topiary sculptures and a plastic pink flamingo on the front lawn. Pushing through the side door of the house, we found ourselves in the Wild Things gift shop, surrounded by stuffed animals, postcards, and paw-shaped note pads. We were greeted by a pair of three-legged house cats—one striped like a tiger, the other with black-and-white spots. A young woman with tan skin and long black hair eventually came out of the back and ushered us through the house to a room with a cash register, a refrigerator with drinks, and small tanks holding snakes and lizards.

The Wild Things exhibits communicated mixed messages about the power to tame animals, to control unruly creatures, and obligations to care for animals.[21] The facility claimed to have "rescued" all of their animals, only ever purchasing animals as companions for lonely residents who might like to breed. Some of these animals, like the Mystery Monkey, had been found roaming in the streets of Florida. Others had once been kept as personal pets—purchased by members of the public from wildlife dealers who import millions of live animals to the United States each year.[22] Rehabilitation programs aimed at reintroducing captive animals into the wild are often designed to instill fear and even hatred of humans.[23] In contrast, the architectures of confinement at Dade City's Wild Things tacitly instilled the opposite: human fear of caged animals.[24]

As the tram approached the Mystery Monkey's enclosure, our guide told us about his visitors from all over the world. But the once-wild monkey was spending most of his time hiding in his nest box, disappointing people from faraway places like Asia and Australia. A Plexiglas screen separated the monkey's small cage from visitors. The whole enclosure was framed by a warning sign: ZOONOTIC Monkey Herpes B. Tossing some peanuts and marshmallows over the screen, the guide said that he was required to "suit up" in full hazmat protective gear every time he cleaned the cage. "The monkey herpes virus can really mess up humans," he said. Herpes B virus (*Cercopithecine herpesvirus* 1; BV) is similar to the herpes simplex viruses of humans. Serious disease due to herpes B virus is rare in macaques, but when transmitted to humans, the virus has a fatality rate greater than 70 percent if not treated promptly. Since 1933, there have been forty-three reported cases of contraction in biomedical or breeding facilities.[25]

Continuing his monologue, our guide said that Cornelius's origin was still something of a mystery. He recounted vague rumors about a wild troop of monkeys that had escaped from the original *Tarzan* movies. "I don't know how many generations ago that might have been," the guide added, "and I don't know how many monkeys are out there, or if they are infected with herpes." Florida officials have known about the presence of these wild monkeys, and their herpes B status, for decades. Richard S. Hopkins, Florida's deputy state epidemiologist, wrote an internal memo on August 15, 1990, stating, "The hazard to visitors of B virus infection from monkey bites or scratches is clearly not zero, but may be small." The memo goes on to remind other officials of hazards posed by alligators, mosquitoes, ticks, and poisonous snakes. "Monkeys are not the only potentially hazardous animals present in Florida parks. . . . It seems likely that the greatest hazard of morbidity and perhaps mortality is from fire ants and other arthropods." There are no reports of the herpes B virus being passed to humans in the wild despite the fact that people and macaques have lived alongside one another in South and Southeast Asia for centuries and have high rates of physical contact at tourist sites.[26]

But our guide's uncertainty about the dangers of simian herpes fed into the mixed messages broadcast by Dade City's Wild Things—about the ability of people to bring unruly elements of nature firmly under control, about the benevolence of owners who rescue animals only to house them in small enclosures. Kitschy aesthetics tried to reinforce conventional messages about taming wild animals, appealing to popular tastes in a way that was easy to market and somewhat fun for visitors to consume.[27] But as these aesthetics failed, other messages emerged. The cheap fencing suggested that the managers of this facility had a limited capacity to separate wild nature from tame culture. Critters confined inside were poised to emerge if given a momentary opening. Departing from Wild Things, we began to study monkeys who had already escaped beyond human attempts to fence them in. Tracing the rumors about monkeys from the *Tarzan* movies to their source, we found surprising multispecies interactions.

THE MONKEYS ON TARZAN'S RIVER

Large groups of free-ranging monkeys, some 130 individuals all told, live in swamplands and riparian woodlands along the Silver River, about two hours north of Dade City's Wild Things. The Mystery Monkey may have originated from the Silver River, since male macaques generally disperse as they approach adulthood, seeking out new groups of unrelated individuals to join. Named for its crystal-clear water, this short spring-fed waterway has a storied history. Silver Springs, an amusement park at the headwaters of the river, began featuring glass-bottom boat tours and encounters with jungle life in the 1890s. In the early twentieth century, when car travel was new and exciting, Silver Springs was a premiere attraction of Florida's "inner frontier"—a wild area alongside densely inhabited cities and towns.[28] Silver Springs' glass-bottom boats, in the words of Wendy Adams King, enabled tourists to experience "beauty, motion, and spectacle" while negotiating "between romantic, scientific, and cinematic visions."[29] Movies like *The Yearling* (1946) and *Creature from the Black Lagoon* (1954) were filmed here along with the 1930s *Tarzan* classics featuring Johnny Weismuller, which have a strong grip on the local and national imagination.[30]

When I took a glass-bottom boat tour in January 2012, the boat captain helped build the mythology of the *Tarzan* movies by weaving in stories about the local population of wild rhesus macaques. The captain offered richer details of the same rumor I heard at Wild Things, claiming that the monkeys were originally brought to Florida as movie extras, alongside Tarzan's famed companion, Cheetah the chimpanzee. Digging through local newspaper archives, I found a more credible story. Four rhesus macaques were introduced to a small island on the Silver River in 1936, according to a local investigative reporter, in hopes of attracting more tourists. Colonel Tooey, "a big heavy fellow with a red face and sandy hair" who operated a jungle cruise boat tour, purchased them from a circus sideshow based in Rochester, New York.[31] After creating Monkey Island in the middle of the Silver River, by dredging a short canal and outfitting the islet with houses and swings, he turned them loose. But, unbeknownst to Colonel (that is his first name, not a military title), rhesus macaques are very capable swimmers. They quickly escaped from the island and began to populate the woodlands of central Florida.[32] Colonel Tooey and the glass-bottom boat captains from Silver Springs began feeding the monkeys as a regular part of their tours, in hopes of keeping some near the river.

Only a handful of people who frequent the Silver River are careful students of monkey behavior. Among them is "John Daniels" (a pseudonym), a former park ranger and proud National Rifle Association member who describes himself with self-deprecating humor. "I'm a redneck," John told us, "but not the kind of redneck that goes shooting things indiscriminately, even if I'm hungry." While we looked for wild monkeys along the Silver River, John paddled his sit-on-top kayak telling us about the rich landscape, a palimpsest with layer upon layer of cultural and natural history. Pointing to a big cypress with a distinctive Y-fork of branches where the original Tarzan, Johnny Weissmuller, once swung, John said that this tree likely sprouted in the fifteenth century, based on the slow-growth habit of this species. He said that the cypress tree has stood as a silent bystander to hundreds of years of human history—as Ocali Indians piled shells into middens, as violent skirmishes erupted with the first white settlers, as local men and goods were channeled to Confederate forces during the Civil War, and as the "colored" beach was established at Paradise Park.

The *Tarzan* movies feature the ape man swinging from this tree by a vine, but it was actually metal fiber cable. By 1986 this "vine" had become rusty and frayed, so John cut it down. "If you'd a tried to use it, you'd have ripped yourself to pieces," he told us. The Tarzan Tree now stands right next to a canal that was dredged at the end of segregation, when the African American beach, Paradise Park, was destroyed.[33] John pointed out

FIGURE 7.1. A scene from the original *Tarzan* movie series featuring Tarzan (Johnny Weissmuller), Jane (Maureen O'Sullivan), and Boy (Johnny Sheffield), with Cheetah the chimp. The famed Tarzan Tree is adjacent to the abandoned site of Paradise Park, the riverside attraction "for colored people" near Ocala, Florida. Photograph by Bettmann/CORBIS.

a couple of exotic palms that you can see in old Paradise Park postcards, a queen palm and another with long spikes at the base of the fronds. After pointing this out, John told us a racist joke that revealed his political sensibilities: "How have you been doing, Tarzan? Well, I'm getting old. My knees are feeble, and I can't swing on the vine like I used to. How's Jane doing? She's got Alzheimer's and doesn't even know who I am anymore. What about Boy? Well, Boy didn't really ever make it. He got into drugs and then went off to South America and is in psychedelic land, but oh well. I'm sorry to hear that. But what about Cheetah? Cheetah moved into the White House."

Chasing after a baby alligator and grabbing it by the tail, John suddenly brought us into a situation where risk, danger, and chance could have easily resulted in an accident. After letting the alligator go, he talked with boyish glee about how he enjoyed escaping the humdrum nature of civilization on the Silver River. The actual violence of grappling with the baby alligator matched the symbolic violence of John's Tarzan joke.

FIGURE 7.2. Cynthia Graham, a local African American historian who is writing a book about Paradise Park, said that white kids and black kids living in a time of segregation both imagined that the grass was greener in different parts of Silver Springs. Paradise Park had a jukebox with music. They had dance parties in a space where whites could not go. The whites had upscale gift shops, exotic animal encounters, and rides. "They could see each other across the river," Graham said. Many African Americans Graham interviewed said, "We had all the fun." Photograph courtesy of Bruce Mozert.

Vigilante lynchings of black Americans happened regularly in Florida up until 1945.[34] A trailer park in Forest, a nearby town that some claim was named for the first Ku Klux Klan grand wizard, Nathan Bedford Forrest, had a "No Niggers Allowed" sign hanging at the entrance when I visited in the late 1990s. Mary Williams, a longtime civil rights activist, later told me that black folks would drive through Forest to the popular destination of Daytona Beach only if they clearly knew the route and if they were expected by someone. Still, Ms. Williams said that the colored beach at Paradise Park was "a source of great joy for my family. The boys would have their bathing suits on in the car. When the door opened, they would just run down into the water. That was the only way you could get in because the water was so cold."

After Paradise Park was bulldozed in the 1960s, marking the end of segregation, Mary Williams's family no longer had a place to swim in the Silver River. The river became dominated by white recreational boaters. Men like John Daniels became free to go wild on Tarzan's own imaginary stomping grounds. Tarzan has long been imagined as a figure who transcended the "stale, mundane and, in a way, unreal" aspects of industrialized life, in the words of Daniel Bradburd. "If modern life made men physically and morally soft, circumscribed and interdependent," according to Bradburd, "Tarzan was the imaginary antidote."[35] In contrast to Dade City's Wild Things, where encounters between animals in cages and visitors in trams reinforced dominant messages about the circumscribed nature of modern life, the Silver River had become a place where white men like John could demonstrate their masculine physical prowess or mastery over untamed creatures. John also offered us glimpses of more subtle forms of wildness. When we encountered the monkeys during this tour, he showed us how they were entangled in social and ecological networks (occasionally involving humans) while keeping their capacity for flight.[36]

Rhesus macaques are attentive students of human behavior and technology. John told us that they would be indifferent to our approach in kayaks but would materialize out of the woods with the sound of crinkling plastic wrappers. Illustrating his point, John quickly produced a small crowd of monkeys just by opening and closing an empty snack wrapper he had on hand. The crowd eventually dispersed since no actual food was on offer. John told us, "I might be in my boat, forty feet back, and as soon as I unzip my cooler, they are gonna look, they are gonna chatter. That's just being conditioned like ole man Pavlov said." He added, "Hear zipper. See food. Come to edge of water. Get reward. Okay. Good behavior." Monkey mothers teach their babies how to live in the world, John said. "Infants riding around on Momma's belly or her back know that every twitch means something. When they get older, and start venturing off on their own, Momma will slap baby's hand, or bark, saying, 'Don't touch that.' When the baby really steps out of line, Big Red the alpha male will intervene and put it back in line."

Even if John Daniels's discourse evoked patriarchal tropes, he nonetheless was attuned to aspects of macaque social behavior (namely policing by the alpha male) as well as the subtleties of macaque cognition and affect.[37] John recounted some monkey behaviors as evidence of higher-order learning—an awareness of the context of their interactions with people. "If you meet one of these monkeys on land, when you are walking and they are up in the trees, and you stare directly at them, they will raise hell and come at you," he said. "But if you are in a boat and you stare at them, they will just sit there on the bank looking back, waiting for a hand-out." Thus they clearly distinguish between kinds of beings in the world: boat people are distinct from land people. Boaters with food are beings of active interest, while boaters with none are of little concern. Macaques understand human ontologies in terms of our actions. Boaters who are feeding are beings that they are actively trying to generate. Monkeys try to transform stingy boaters into feeding boaters by cooing and even staring.

BECOMING VULNERABLE

In macaque societies, it is impolite to stare. Eye contact is threatening. "Direct eye to eye contact means, 'I challenge you,' 'I am boss around here,' or it can mean 'stop that,'" according to Bob Gottschalk, a retired engineer from Iowa, who studies the Silver River monkeys as a citizen scientist. "In most cases eye contact between humans and monkeys means 'I chal-lenge you.' Of course, that wouldn't be wise," he wrote in a report about the monkeys for local officials. "The wise action is to look away, which means 'I don't want any trouble,' or 'I accept the fact that you are the boss.' "Mouth open, corners of the mouth forward, and ears back are signs of a serious threat," he adds.[38] Primatologists, who have authored extensive inventories of macaque behaviors called ethograms, generally refer to this as an "open mouth threat." Standard inventories of aggressive behavior also include "branch shaking," the "yawn threat," and "head bob threat." More subtle behaviors include the "eyelid flash," the "silent bared-teeth display" or "fear grimace," and the "lip smack."[39]

In collaboration with Tiffany Wade and Erin Riley, who were conduct-ing a parallel study of macaque foraging ecology, I considered the power dynamics in multispecies worlds. Rather than using standard ethological techniques for quantifying the frequency of predetermined behaviors, I used descriptive methods to characterize the dynamics of human-monkey interactions on the Silver River. I found macaques using impolite gestures from their own social worlds to generate behaviors they desired from humans. In exploring one key question—cui bono, who benefits, when species meet?—I found evidence of power plays in parallel social worlds. Power dynamics in macaque societies have been extensively studied by primatologists.[40] Female rhesus macaques obtain rank based on the posi-tion of their mother within the matriarchy, with each new daughter get-ting a higher rank than her older sister.[41] Males generally join new groups of unrelated individuals as adults and constantly jockey for position. They usually follow a "seniority rule" where they rise in rank as other males

FIGURE 7.3. It is impolite to stare. Photograph by Eben Kirksey.

leave or die. The alpha male, who is dominant over all other members of the group, usually does not hold this rank very long. The top-ranking female is dominant over all other females and males (with the exception of the alpha male). Nested hierarchies produce one counterintuitive result: the third-highest ranking member of a rhesus macaque society is the alpha female's youngest daughter.[42]

While paddling on the Silver River with Bob Gottschalk, who has been studying these monkeys for over five years as a citizen scientist, I learned about the history of political dynamics within four distinct rhesus macaque groups: Troop A, Troop B, Troop D, and Troop E. Bob generally came out to the river every Tuesday in winter months, developing knowledge of the troops as he recorded their movements, made general behavioral observations, and studied the personality of individual animals. In January 2011, when we first ventured onto the Silver River together, Troop E was the biggest mob of macaques, with two males (whom Bob had named Scar and Broken Leg), twelve females (whom he left anonymous), and upward of thirty-five juveniles. "Broken Leg has been with this troop for a long time," Bob said. "He is allowed to be there because he is good protection for the troop. He is no threat to Scar. With that leg, he is no threat to another healthy male." Bob told me that Scar tended to monopolize encounters with boaters, usually securing most of the prized food for himself. The answer to the question of *cui bono*, who benefits from these encounters, initially seemed simple. With focused field research, my interdisciplinary team began to find more complex answers.

Returning to the Silver River a year later, we used the knowledge of Bob Gottschalk and John Daniels to help frame our ethnographic and ethological fieldwork.[43] Rules about monkey-human interactions formulated by Bob and John generally held true, even if some particular events led us to note exceptions. Our research was also guided by rules formulated by the government. According to the official policy of Silver River State Park, which began managing this waterway in 1985, feeding wildlife is strictly prohibited. In actual practice, many visitors give the monkeys food. Our permit application, to the Florida Department of Environmental Protection, proposed to "examine the patterns of human-macaque interactions along the river." Our stated goal was to study "the nature of interactions (contact or non-contact), responses to the interaction by those involved, rates of food provisioning by people, the percent contribution of human foods to the macaque diet, and how different transportation technologies (e.g., kayaks, canoes, motor boats) structure interactions." In other words, we took careful notes as people bent and broke the rules.

Sometimes people seek "to tame wild animals by feeding them," writes Molly Mullin, "or at least to change their behavior in significant ways, thus encouraging connections between domestication and control, wildness and freedom."[44] On the Silver River, we found some people feeding the monkeys in patterns that allowed youngsters to gain temporary freedom from macaque dominance hierarchies. We also found recreational boaters who subscribed to libertarian values, who were seeking freedom from government regulations hemming them in. One Sunday afternoon, as a family in a blue motorboat fed grapes to Troop E, we observed particularly interesting dynamics as human norms about feeding wildlife interacted with macaque rules about access to food. Monkeys were jumping among the trees, clambering around the exposed roots of red maples and cypress trees, as grapes sailed across the river. Three adult females were at the edge of the water in among reeds and tree roots. Almost all of the grapes were falling into the water. In my field notes I wrote:

> Scar, the alpha male, lurked behind a tree, looking over a small gaggle of yearlings gathered on the riverbank. Dad, who was wearing a blue jacket, sunglasses, and a black baseball cap, threw grapes while dangling his legs over the side of the boat. Aiming at a small cluster of monkeys on the other side of the river, he said, "Hope they see the food and jump in." Jake, the oldest son at around eleven years old, followed suit and started chucking grapes too. As the females and yearlings sifted through the reeds and the mucky bottom for grapes Mary, an off-duty ranger from the state park, approached in a kayak, paddling upstream.
>
> "Please don't throw things at them," Mary said as she paddled up alongside the blue motorboat.
>
> Dad replied, "We're feeding them grapes."
>
> "Please don't feed the monkeys," Mary continued.
>
> "Why," asks Dad, "not good?"
>
> "Ask the researchers," she implored.
>
> Mom looked at us. We shrugged.
>
> "They don't care," said Mom.
>
> "The state park does," Mary fired back. "It's kind of against our rules."

Suddenly we became implicated in a dispute between the subjects of our study and park staff. While mixed up in this encounter between recreational boaters and an off-duty ranger, we were momentarily drawn into a power struggle, with our tacit approval of the feeding impacting the very dynamics we were studying. Macaques often draw people into their own power struggles, according to John Knight, whose article "Feeding Mr. Monkey" describes tourist encounters with Japanese macaques at monkey mountain theme parks. Knight suggests that dyadic interactions between pairs of monkeys are in fact based on a "triadic sociality." The presence or absence of a third party crucially affects the interactions of dyadic pairs. Knight documents how humans are often brought into these dynamics of power. "Monkey behaviour towards visitors is likely to differ depending on whether keepers are nearby or not," writes Knight. "Where keepers are out of sight, a monkey may well act more boldly than where keepers are present. This is because the monkey knows that aggression against the visitor in front of the keeper could lead to retaliation by the keeper (who usually carries a catapult or a stick to use against the monkeys)."[45] As we were drawn into a triad, a power struggle between the boaters and the off-duty ranger, subtle power dynamics were also playing out in the macaques' social world.

> In the background, a juvenile belly flopped into the water, dive-bombing a grape it spotted on the bottom of the river. Another yearling cooed. He was also studying the situation, carefully watching the dynamics among the people and within his own social world, hoping to score a tasty grape. As Scar looked away, the youngster looked right at Jake, the boy who had been throwing grapes. Breaching polite norms of macaque society, the young monkey returned Jake's gaze and softly cooed again—begging, or perhaps daring, him to make another toss, reaching across gulfs of biology and meaning.
>
> Looking down at her son, and then at Mary, Mom asked, "Well, why? Why shouldn't we feed the monkeys?"
>
> "It's the food chain," Mary said. "If you are feeding the monkeys then you are feeding the fish, and then pretty soon the alligator will get it, and then when an alligator sees a human . . ."

After this tense interchange, Mary continued upstream, trying to enjoy her off-duty day on the river. Mary was trying to make vulnerabilities visible in an interspecies encounter.[46] The specter of another agent, the alligator, illustrated that this shared bubble of happiness was fragile, like glass, and could shatter at any moment.[47] American alligators grow to an enormous size on the Silver River, and we often found ourselves anxiously looking around whenever the monkeys made an alarm bark, a sound reminiscent of a screechy dog, and gestured to the depths of the water nearby. Alligators occasionally kill and eat baby monkeys, according to park staff who witnessed two separate incidents. While humans have occasionally been killed by alligators in Florida, with twenty-two fatal attacks since 1948, crocodilians are being killed by the thousands. In 2006, Florida's Fish and Wildlife Agency killed a record 11,664 alligators after an increase in complaints about "nuisance" animals.[48] After Mary rounded a bend in the river, as yearlings continued to dive-bomb grapes, we noted facial expressions hinting at mixed emotions and ambivalent feelings in people and monkeys. Unequal risks were clearly at play—with small monkeys in the water, humans safely in boats, and nearby alligators in danger of becoming a nuisance. Jake (the eldest son) looked to his parents for guidance, and then continued to toss grapes toward the monkeys, escaping sanction as he continued to break official park rules. Scar circled around the cypress tree roots seeking access to the occasional grape landing on firm ground, but he did not approach the river's edge. Zig-zag, an adult female named for the shape of her disfigured tail, appeared on the scene, chasing off an infant with open-mouth threats and lunges. As a grape sailed over Zig-zag's head, into the reeds, she looked over her shoulder. A yearling ran in, intent on snagging the grape. But Scar ambled over, his tail cocked high in a dominant gesture, and picked it up.[49]

"Monkeys do not like to swim," John Daniels later said. "They are very wary, very cautious, aware of what is around them when they come down to the water. You would think they are not cautious by the way they come down to people feeding them," John continued. "They look like they are having a great time, frolicking and jumping." The grapes falling in the water had created a situation of heightened risk, where the usual social rules about the distribution of food did not apply. While Scar behaved with caution near the water, younger macaques were quick to exploit the opportunities opened up by a dangerous situation they perhaps lacked the experience to understand. Ultimately, this encounter (and most of the other interactions I observed on the Silver River) benefited the monkeys and became a memorable experience for the boaters. It followed the plot of a comedy—it turned out happily for all involved. Monkeys received food; no one was hurt; no punitive fines were issued by park staff. Proximity to humans and shadowy specters in the water enabled young macaques and a young boy to become a little wild—to become a bit unruly, to find autonomy and freedom in a space yielded to them by domineering

FIGURE 7.4. Pontoon boaters and a dominant female macaque.
Photograph by Eben Kirksey.

elders and state officials. In becoming wild, macaques and humans were inducing each other to disrupt power hierarchies. While participants in this particular event were relatively restrained, other encounters became increasingly intense, with redoubling loops of affects and emotions. Infectious feelings began to quickly jump between people and monkeys.[50] Wild encounters began to spin out of control.

BECOMING EXCITED

As the buzz of a motorboat sent vibrations down the river, several adult female macaques sat at attention and scanned the distance. This was a year after our initial joint field trip to Florida. The group with Scar and Broken Leg had grown to fifty-five macaques, and they were spread out along the south side of the river. Yearlings were play fighting and chasing one another from tree to tree. Scar was being elusive, resting in the shade of a saw palmetto, taking a break from eating the buds of Carolina ash (*Fraxinus caroliniana*). Fifteen monkeys, led by King Philip and Queen Isabella (Bob's names for Troop A), were scattered on the river's north shore among pumpkin ash (*Fraxinus profunda*) and red maple (*Acer rubrum*). As the buzz from this motorboat reverberated in the water, in our bodies, and through the groups of macaques, we shared a sense of heightened excitement and heightened risk.[51]

> A red motorboat zipped around the upstream bend of the river, driven by a man in a cowboy hat and board shorts. Delighted shouts rang out.

> "Oooh, look, a baby monkey!" Infant monkeys cooed. As the rumbling of the motor deepened, as the emotion in the boaters' voices reached a peak of intensity, affects became contagious, jumping across the species interface.

> Macaques stopped eating, grooming, and resting. They hustled to the river's edge en masse, with the youngsters excitedly cooing for food. As a bald man started throwing large chunks of banana, swarms of monkeys started running, on both sides of the river, toward the boat. Scar ambled toward the front, becoming an assertive presence. Squabbles erupted as he, and other bigger monkeys, moved in to nab choice bits of banana from yearlings. Contagious affects again jumped between species, infecting new groups of people as they arrived on the scene.

> Two kayaks paddled up, plus another motor boat—a 115-horsepower steel skiff. One of the kayaks, carrying a dad and a boy, about five years old, paddled toward the monkeys on the north shore.

> "Look at the baby one!" shouted the boy.

> "Want to go say hi?" asked Dad.

> "No, no, no! I don't wanna go near the monkeys!" replied the son as the father began to back paddle.

The pitched intensity of interaction grew, exacerbating a situation of mutual vulnerability.[52] As humans and monkeys became wild together, as delight and fear became highly contagious, the threat of violence emerged amid the breakdown of established hierarchies and boundaries. Moments after the kayak backed away from the monkeys—"No, no, no! I don't wanna go near the monkeys!"—Black Cajun, a motorboat louder than all the rest, suddenly appeared around the bend. People on this boat tossed Cheez-Its and nuts toward King Philip and Queen Isabella's group on the north river bank. After a momentary lull, a pontoon boat suddenly approached with a large intergenerational group aboard. As soon as they spotted Troop E, with Ester, Scar, and Broken Leg, on the southern bank of the river, they started tossing jalapeño crackers at them. After exhausting their bag of crackers, they began throwing whole fruit.

Oranges and apples made it easily to shore. Monkeys rushed out from behind logs and shimmied down trees. A middle-aged man on the boat tossed an orange to an adult female. She caught it with two hands and immediately ran off.

"Good catch, bubba!" said the man.

A subadult male was the next to get an orange, but did not get away, like the female before him. Once he acquired the orange, the adult female closest to him grunted in opposition and lunged toward him. He turned and ran inland, orange in hand.

Four higher-ranking females recognized this young male's transgression and started to put him in his place. Grunting, they chased him up and back down trees. The grunting became louder, more intense, and more rapid as they all hit the ground and started running, hidden from view. Amid the bedlam, the young male emitted a sharp cry, probably in response to a bite. The pontoon boaters stopped tossing fruit and stared blankly while the grunts and cries continued.

"Hey! Gawd!" the man screamed. "I'm the man! Hey! Go home!" His family chimed in with chimpanzee sounds: "Oo ou ah ah eee eee eee!"[53]

Wild and unruly behavior by the boaters exacerbated existing social tensions within this macaque group. A young male made a transgression, seizing an opportunity on the fly, only to be immediately punished by dominant members of a matriarchal society. Following the climax of this interaction, as the pontoon boat motored off downriver, the macaque group quickly returned to a steady state, where prevailing order was reestablished after a momentary rupture.[54] Outbreaks of macaque-on-macaque violence were not uncommon during our fieldwork, which took place at the height of breeding season—a period when male and female rhesus macaques showed facial color changes and had genital swellings. Other studies have linked the macaque breeding season with increased muscle mass and weight gain for subadult and adult males as well as an overall increase in rates of aggression.[55] Even as human food and contagious affects exacerbated internal macaque dynamics to occasionally produce open violence, practices of care and fleeting alliances mitigated situations of risk and vulnerability.

BECOMING POLITE

While some visitors seem bent on deliberately inciting conflict, others, like Bob Gottschalk, have carefully thought about the norms of interspecies tact, politeness, and regard.[56] When Bob is paddling on the Silver River, conducting research approved by Florida's Department of Environmental Protection as a citizen scientist, he rarely intervenes when other boaters feed the monkeys. Instead, he tries to teach them how to become polite, offering informal lessons on social rules and norms governing macaque society. He usually advises people to feed only the alpha male or the high-ranking female monkeys. Just feeding the youngsters will invariably generate conflict, Bob says. For people who are intent on feeding the babies, he recommends a scattershot method—distributing lots of food throughout the group in rapid-fire style. Bob works to encourage fellow members of the boating community to cultivate tactful relationships with wild animals. Creating spaces where quiet forms of wildness might flourish, Bob supports the monkeys' autonomy—their freedom to approach and withdraw from humans.[57]

Careful attention to polite social cues enabled Bob to become close to one monkey group in particular. He told me about his relationships with individual monkeys, telling stories that offer evidence of mutual recognition and regard. Bob felt closest to a monkey he called King Louis, formerly the alpha male of Troop A. Bob said that he would carefully watch King Louis's gestures, gently navigating a cautious approach. If the king was in a good mood, Bob would get out of his kayak sometimes and sit up on the river bank, next to him. The pair would just hang out in the same space, with Bob occasionally tossing a peanut to his monkey friend. Mark Waiwada, a volunteer who polices the river in a kayak while wearing a tan button-up uniform emblazoned with a white-and-blue Florida State Park logo, has a very different opinion about interspecies politeness and tact. Politeness for Mark involves keeping a respectful distance. "Monkeys have entertainment value," Mark said. "If you don't feed 'em, they aren't a danger. My great fear is that someone will get bitten and then we will have to kill that monkey, that entertaining creature."

Politeness on the Silver River is also mediated by boating technologies. Traci Warkentin has written about the ethical affordances built into the architectures structuring human-dolphin encounters. Walled barriers at SeaWorld mediate coercive feeding and touching. Pools where one can swim with the dolphins usually afford the captive animals no opportunities to escape. Warkentin suggests that open oceans, where dolphins can approach and withdraw from humans at will, make for the most ethical interspecies contacts.[58] Interspecies relations on the Silver River are certainly structured by material asymmetries, with humans possessing an excess of tasty food. Still, it is clear that boats generally afford more ethical and comfortable encounters than the other sites of human-monkey interaction in diverse parts of the globe.

Human encounters with macaques in Singapore are mediated by cars. People routinely pull up along roads that are frequented by monkeys and toss food out of their windows. This means that people are protected from the monkeys by the architecture of their cars, but that the monkeys are exposed to the danger of being hit.[59] At Balinese temples and a roadside stop in Gibraltar, encounters are relatively unconstrained—with people on foot. Monkeys sometimes climb up the bodies of the tourists—often stealing sunglasses and cameras, demanding gifts of food in return. While these encounters offer a more level playing field, they are also marked by heightened risk.[60] On the Silver River, I heard lively stories of monkeys jumping onto boats to grab food from taunting children, but I never observed a single contact encounter on the Silver River. Monkeys usually quickly retreat up the trees that line the riverbanks or run away from the water's edge if a person gets too close or is threatening. People usually quickly back their boats away from the shore if their own comfort zone is invaded.

Monkey bites are relatively infrequent at other sites around the world where humans feed macaques—even at places like Gibraltar or Bali where they regularly climb up on tourists.[61] Agustin Fuentes describes a young Swiss tourist in Bali who made the mistake of snatching a macaque infant from its mother, precipitating the worst attack he had observed in twenty years of observing monkey-human interactions. The tourist was immediately attacked by five adult females and one adult male and ended up hospitalized with over 140 stitches.[62] By contrast, monkey bites are relatively rare in Florida. A study commissioned by the state of Florida reported "a total of 23 monkey-human incidents resulting in human injury and 8 incidents where no injury was recorded during the period 1977 through 1984. Of the 23 incidents, 6 were cases reported to the Marion County Health Department. All 6 cases were provoked."[63]

Provoked monkey bites are often the result of rowdy human behavior. Wild men (and sometimes wild women) regularly engage in mimetic dances with the monkeys on the Silver River, looking at them as mirrors and windows revealing their own imagined savage nature.[64] Flirting with risky situations, aggressively trying to subjugate animals or show them who was the boss, some boaters tried to demonstrate their ability to tame nature.[65] Like the out-of-control settlers who brought wildness to the Australian outback, these unruly visitors generated chaos where there was once stable quietness.[66] Recklessly taunting the monkeys with food, impolite people entered spaces of shared vulnerability—occasionally becoming injured themselves. As I spent more time on the Silver River, I learned about an earlier historical moment when contagious unruliness seemed to spiral out of control, when vulnerabilities shared by wild people and monkeys became increasingly asymmetrical.[67] State agents tried to step in, trapping and culling monkey populations in an attempt to bring a wild situation under control.

CATCHING THE MONKEYS

In the 1980s, the Fish and Wildlife Conservation Commission began promoting a plan to eliminate wild rhesus macaques in Florida—claiming that they harmed native ecosystems and posed public health risks. The Audubon Society fueled the fire with allegations that the monkeys were eating the eggs of native bird species.[68] The allegations by established environmental organizations were contested by Tish Hennessey, a native of upstate New York who retired to Florida because of a chronic lung condition. Tish was a key local organizer who came to the monkeys' defense. I learned about the history of Tish's animal activism while visiting her house at All Creatures Sanctuary, a compound of mobile homes and jury-rigged fences where she lives amid an incessant chorus of squawks, barks, and snorts from a diversity of rescued birds, dogs, and other exotic animals. Elan Abrell was conducting his own field research here within a small labyrinth of enclosures for coyotes, foxes, pigs, and deer. Opportunistic vultures hop among the sanctuary residents, pilfering scraps of food. In the 1980s, Tish helped found an organization called Friends of the Silver River Monkeys. As Florida state officials began trapping the monkeys and selling them to laboratories for biomedical research, legions of animal rights activists—including People for the Ethical Treatment of Animals—joined her cause. She mobilized the citizens of Florida to protect the rights of macaques to live in the wild.

After helping gather some fifteen thousand signatures on a petition to Florida state legislators, Tish joined a caravan in 1993 up to the Capitol Building in Tallahassee to present it during a hearing about the status of the monkeys. She was in a wheelchair at the time, since her lung condition was worsening. When she arrived in the hearing room, the legislators were acting like monkeys, according to Tish. They were joking among themselves, scratching their heads and armpits, and carrying on. As she was wheeled up to a microphone, Tish dropped the box with the signed petitions on the floor. The box made a resounding thud, stopping all of the antics in the room. Tish's political organizing and dramatic performance that day won the monkeys a stay of execution. From that day forward she made an avowed commitment to watch over and protect the monkeys. Since her own health was poor, and she could not go paddling on the river herself, she quietly ran a network of volunteers out of her mobile home. These monkey lovers had a covert program of surveillance and stood ready to spring into overt action on a moment's notice.

After the political standoff in 1993, Tish Hennessey began to suspect that the culling of monkey populations was continuing clandestinely. Suspicions of government secrecy also surround plans by other parks to kill introduced animals elsewhere in the world, like the wild donkey populations of the U.S. Virgin Islands. Residents "claim the Park Service continued to cull donkeys outside of the public's eye," according to Crystal Fortwangler, even though "the Park Service adamantly denies this."[69] Tish had long suspected that Florida's monkeys were being trapped near the state park, despite denials by officials. Her suspicions were confirmed in January 2012 when a story broke in a local newspaper, the *Ocala Star-Banner*. "Catching, Selling Silver River Monkeys Is Lucrative," the newspaper headline screamed. Scott Cheslak, a veteran monkey breeder who once oversaw U.S. government macaque colonies on a sea island in South Carolina, had been quietly trapping wild monkeys in Florida for over a decade. Over the years Cheslak trapped more than seven hundred monkeys.[70] Medical laboratories will pay upward of $15,000 apiece for Rhesus monkeys, according to newspapers. But when I reached Cheslak by phone in July 2015 he said that the figure of $15,000 was inflated. He said that the newspaper stories contained many inaccuracies, but he would not disclose the sale price of the monkeys or details about the trapping.[71]

According to one longtime Ocala resident who occasionally helps Cheslak out, he usually arrives in town without telling anyone. "It's like an undercover sting operation," said this informant, a woman who insisted on meeting me late one night at the Cabin Pub, a bar frequented by hunters in the Ocala National Forest. "He hires a driver as well as a river guide who helps locate and trap monkeys on the river." Once everything is in place, Cheslak goes out after dark to strategically place traps along monkey trails and in "choke points" where the ridge line forces the monkeys close to the river. Early the next morning he goes out to check the traps, armed with a dart gun and sugar candy laced with drugs. When he is out on the

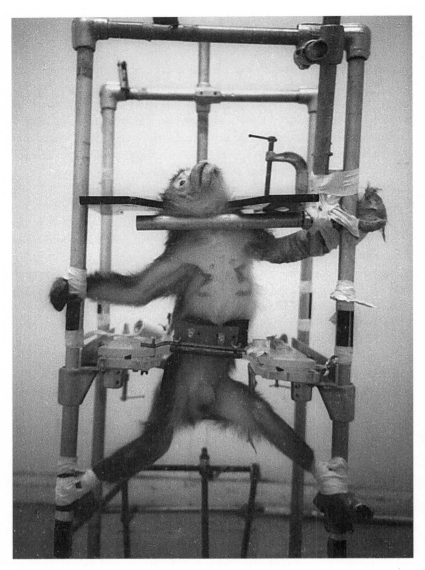

FIGURE 7.5. Animal rights campaigns in the 1980s began circulating images of monkeys subjected to painful treatment in laboratories. Local efforts by All Creatures Sanctuary were linked with campaigns throughout Florida and the United States by activist newsletters sent through the mail as well as faxes about urgent issues. Photograph by Alex Pacheco.

river, Cheslak switches up his boats and his attire, our informant said, to avoid detection by both the monkeys and Tish Hennessey's network of undercover monkey lovers. Burlap bags are used to hold the monkeys in the boat, and Cheslak gives them an injection if they start to wake up. Along the river he has set up a number of hideouts, blinds where he waits to dart the monkeys if they wander by. "Scott is very private," said my anonymous source. "I don't work with him all the time, only when he asks me to."

As I sat and drank with this seasoned tracker, she revealed a deep sense of ambivalence about her work with Cheslak. She told me about watching "a momma who had just lost a son or a daughter." When they lose a family member, "that pisses 'em off. They don't like that." She initially implied that the monkeys quickly recover, saying, "They seem to accept it and move on." But in the next breath, without any prompting from me, she said, "I don't think that they ever get over mourning, to tell you the truth. You know that occurs with gorillas, the higher primates, but even these little guys, hmmm. You get to see what you would call 'sadness.'" She went on to describe a particularly sad incident. Once she and Cheslak caught a mother monkey in a hog trap, and the baby refused to leave. "It didn't know any better," she said. "Instead of scrambling up a tree, or running away, it just stood there, tugging at the cage, trying to get at its mother. A hawk was attracted by all this commotion." As Cheslak and my informant stood there watching, the hawk swooped in and killed the baby, dragging it off to eat. While the mother jumped up and down inside the cage, "the whole troop went ape shit."

Cheslak's sometime helper says that the monkeys would be "impossible to eradicate" if state officials ever earnestly tried again. "If you start trying to dart them all, you might get lucky and take out the alpha male and the alpha female, but then the group would probably disintegrate and scatter. They would turn into ghosts," she said. "They would silently flit away in the night." According to a study commissioned by the state of Florida in 1994, "The proliferation of rhesus monkeys has resulted in numerous reports of them outside the Silver Spring Attraction property, both along the Oklawaha River and in the Ocala National Forest. Additional sightings of rhesus macaques have been reported in Pierson, 70 miles by river from Silver Springs, Eureka, 25 miles from Silver Springs, and Deland, over 100 miles from Silver River."[72] Rumors of monkey sightings near Florida's border with Georgia were circulating by 2013. In practical terms, wild monkeys simply cannot be removed from Florida. Like foxes in Australia, which are undesirable for many people, "they cannot simply be scraped off the top of 'real' native ecologies."[73]

Sally Leib took the helm as Silver River State Park manager in 2012, just before the controversy about Cheslak's trapping was reported in local newspapers. "We struggle with the park service mission," Leib told a reporter. "We know people like to see the monkeys, but we know they don't belong here."[74] A spokesperson for the Fish and Wildlife Conservation Commission made a stronger statement: "Monkeys are an exotic species and don't belong roaming loose in Florida."[75] White settler colonists around the globe have scripted countless wild animals into the story of invasive species, a familiar plot demonizing migrants.[76] Classifying a critter as exotic or invasive seems to offer a "simple mapping of natural versus cultural agency," in the words of Stefan Helmreich. "The category *invasive* also requires a further step: social judgment of harm."[77] Grassroots political activists who loved the presence of wild monkeys in Florida pushed officials to take a step in the opposite direction. They worked to pass a collective social judgment of harmlessness.

Following public outcry about Cheslak's trapping, Florida's Department of Environmental Protection eventually announced that trappers would be temporarily banned from lands under their jurisdiction. Amid this controversy, Bob Gottschalk worked to domesticate the wildness of Silver River's monkeys in an idiom that Florida state officials could accept and understand. In a twenty-page report for Florida's Department of Environmental Protection, a summary of his years of field research, Bob concluded, "Having been part of the ecosystem for over three-quarters of a century they have become intrinsic to that ecosystem. They eat non-native plants, they help to disperse seeds of some plants. . . . Though rhesus monkeys are not an indigenous species they are, by virtue of where they are born, native to the land they inhabit." During an appearance on the local Fox television station, Bob similarly argued, "These monkeys appear to be neutral to the ecosystem. They are not taking anything out of it. They are not adding anything to it. They are basically neutral."[78]

It would be easy to quibble with Bob's claims on Fox TV—he did not specifically study how the monkeys were impacting plant communities. There is a vast literature in ecology about how long-term selective browsing by herbivores changes the density of plants as well as the composition of the forest canopy.[79] Competition with other herbivores—potential niche overlap with esteemed mammals like deer as well as unloved others like insects[80]—was also beyond the purview of Bob's study and my own research. Future studies of the Silver River should aim to answer one key question—cui bono, who benefits, when species meet?—in terms of community ecology. Still, Bob's considered opinion about the monkeys being ecologically neutral is worth pondering. In other parts of the world, introduced macaques have become the poster children of conservation initiatives. Macaques that were imported to the Togean Islands of Indonesia in the 1920s were once regarded as "feral" members of a "bottle-necked hybrid swarm." During intense taxonomic debates among Indonesian conservationists and international biologists in the 1990s, the status of these primates shifted from exotic to endemic. Now conservation biologists assert that Togean macaques play a "keystone" role in local forest ecosystems.[81]

Many Florida residents, who are relatively recent migrants themselves, understand the belonging of monkeys in the context of their own family histories of relocation and settlement.[82] Diverse people are finding new ways to get along with one another and other species amid ongoing immigrations and the bad memory associated with earlier frontiers of white settler colonialism.[83] Questions relating to historical representation and belonging on the Silver River were given renewed attention by Florida state officials as the amusement park at the headwaters fell on hard times with the economic downturn. Silver Springs, a tourist trap of yesteryear, had lost its place in the imagination of the contemporary United States as films like The Creature from the Black Lagoon and the Tarzan series faded from collective memory. Palace Entertainment, the company running the attraction, informed the State of Florida that they would like to withdraw from their lease. Closing the attraction jeopardized the livelihood of glass-bottom boat captains, high school kids who worked summers at an attraction called Wild Waters, and people who took care of Silver Springs' diverse population of captive animals—crocodiles, cougars, gibbons, Kodiak bears, and multiple species of poisonous snakes. These constituents came together with state officials on January 14, 2013, for a public meeting at Ocala's Vanguard High School to discuss the fate of Silver Springs.

Officials from the state capital, Tallahassee, opened up the meeting at Vanguard High by presenting their plans for Silver Springs with a PowerPoint presentation. The 186-page document guiding this public consultation, "Silver River State Park Unit Management Plan," outlined the state's proposal for "exotic species removal," including the monkeys as well as the animals being kept in cages at the Silver Springs amusement park. Tearful zookeepers, who were coming to terms with the fact that their jobs were ending, approached the microphone during a Q&A session—asking if the creatures in their care would be euthanized or transferred to other facilities. Avoiding this moment of tension, state officials replied that the details would be worked out later. Cynthia Graham, a black historian who is researching a book about the colored-only Paradise Park, also stood up to the microphone to address the audience of some 250 people, who were almost all white. She noted that Silver River State Park had hitherto failed to recognize the presence of Paradise Park with signage.[84] Graham politely asked the state officials if they would acknowledge the local history of segregation as they developed new signs and interpretive exhibits in the state park.[85] As audience members gave Graham a round of applause, the officials quickly vowed to include Paradise Park in their plans for future cultural exhibits.

The fate of the wild monkeys on the Silver River was not discussed openly at the Vanguard High School public meeting. In the months after this meeting, as zookeepers from Silver Springs worked to find homes for the 270 exotic animals in their facility, monkey-loving activists renewed their campaign to ban trapping.[86] The Animal Rights Foundation of Florida initiated a Change.org petition addressed to three top Florida officials, saying, "These wild monkeys deserve to live free from confinement and abuse." The petition attracted close to two thousand signatures and comments from all over the country. "Most of us are 'non-natives,'" wrote Kate Kenner of Jamaica Plain, Massachusetts. "Wild and FREE should remain that way!! God's intention!!!" added Pamela McCucheon of Yorktown, Virginia. This petition culminated in a public protest on October 1, 2013, the day when the expanded and renamed Silver Springs State Park first opened its doors. Weeks later, on October 25, Florida officials made the temporary ban on monkey trapping permanent.[87]

QUIET WILDNESS

With the opening of the new Silver Springs State Park facilities, traffic of Fish and Wildlife officers on the river increased. Visions of the Silver River as the Wild West, where one could don a cowboy hat and rip through the water at high speed in a motorboat, began to be curtailed by state agents who desired to see this park become quiet country.[88] Nature lovers like Mark Waiwada and monkey advocates like Bob Gottschalk continued to encourage Florida officials to embrace new forms of wildness, to experiment with novel approaches to the care of Florida's damaged country amid long legacies of violence and ecological change. Out on the river, Mark and Bob still tactfully guide interspecies encounters—nurturing wonder about these animals as their social and ecological networks periodically overlap with those of humans.[89] Intervening in situations that threaten to spiral into violence, these committed volunteers ensure that the monkeys retain the capacity for flight and the boaters do not approach too closely. Beyond the park boundaries, a multitude of activists and community leaders stand ever ready to protect the rights of monkeys to live beyond the realms of human supervision and control. Like donkeys, which now wander freely in protected parklands of the U.S. Virgin Islands, the Silver River monkeys are now "wild in the sense that they are unchecked, free to roam."[90]

Taking an anthropocentric and macaque-centric view, it is relatively easy to answer the question of who benefits from this quiet form of wildness on the Silver River. Despite occasional situations of risk, vulnerability, and open violence, net benefits have emerged at the intersection of macaque and human social worlds. In an instrumental sense, macaques come away from these encounters with tasty foods, while people associated with the tourism industry and the state park earn economic revenue. Visitors (and some monkeys) who participate in these interactions do not necessarily derive instrumental value from them, but instead are drawn in by "encounter value."[91] Values reinforced by the tame encounters at Dade City's Wild Things are very different from the possibilities for freedom that have emerged with wild face-to-face encounters between different primate species on the Silver River.

Making careful ethical and theoretical distinctions, it is possible to embrace some forms of wildness while opposing others. Deleuze and Guattari betray an atavistic longing to become "pack or affect animals."[92] Wild men (and women) who use the monkeys as mirrors and windows into their imagined animal nature reveal similar atavistic desires.[93] While rejecting the wildness of colonialism, and of the unrestrained economic systems where monkeys are bought and sold in a biomedical marketplace, it is still possible to support the capacity of these animals for flight and their ability to raise families.[94] The presence of wild monkeys on the Silver River sits uneasily with plans to restore past environments, an imagined state of nature that might never have existed. Flying in the face of long-term human agendas, forms of life are blossoming in wild profusion on the Silver River.[95] Transformative encounters are generating new entangled modes of coexistence. Looking toward the future with curiosity, I found novel ecological associations emerging in Florida as diverse life forms pushed beyond human attempts to fence them in.

CHAPTER EIGHT

MULTISPECIES FAMILIES

...

Consumer values ascribing worth to novelty result in the importation of several million animals each year to the United States.[1] "As a multi-billion dollar a year industry, the global illegal wildlife trade is second only to narcotics," according to an advocacy organization called EcoHealth Alliance.[2] Alongside this black market is a regulated marketplace for wildlife importers and breeders. Florida stands out in the legal landscape of the United States as a place where exotic animals live within a laissez-faire economic environment that permits their existence as property within the moral milieu of the human home.[3] "We allow just about anything as a pet as long as the owner obtains a proper license and proper cages," said one official with Florida's Fish and Wildlife Conservation Commission. "Our agency is not anti–exotic animal. We support responsible ownership. And we are simply not going to spend billions to kill everything that escapes." These policies are generating multispecies families where animals are regarded less as "cocitizens" than as liminal beings who can suddenly be demoted to the status of commodities or things.[4]

Family values in many multispecies households mix notions of kinship with capitalist principles of production and reproduction. Individual animals might be addressed with the endearment "my baby" and then treated as breeding stock or a commodity to be exchanged on the open market. Exchange value and use value were classically part of Karl Marx's account of capital extraction and accumulation.[5] Donna Haraway has added a surprising twist to Marx's classic story of capital with the notion of transspecies "encounter value." Lively capital, by Haraway's reckoning, can generate hopeful coalescences where "commerce and consciousness, evolution and bioengineering, and ethics and utilities are all at play."[6] Studying the home economics of multispecies households in Florida, I found that the bodies of animals are often possessed and then abandoned by the ghostly spirit of capitalism.[7] In a phrase, I found that some kinds of pets are treated as "flexible persons."

The notion of flexible personhood refers to how animals are often temporarily incorporated into the familial sphere, folded into relationships involving love and care, only to be abandoned amid major life changes. Dafna Shir-Vertesh, who studied human relationships with dogs in Israel, coined this term to understand how animals are shuttled among moral spheres where they enjoy different rights and privileges. Pets are often loved, incorporated into human lives, and addressed with kinship terms. But they can be demoted at any moment, moved outside of the home and the family, as household income or personal circumstances shift.[8] "Particularly the birth of a human child," writes Shir-Vertesh, "challenges the pet-as-baby human-animal bond and may render void the animal's loving presence as prechild, semichild, or child substitute."[9] Building on Shir-Vertesh's critical insights, I began to study how animal-children figure into human strategies for dealing with periods of rapid change, flux, and uncertainty.[10]

Flexible strategies for accumulating and selling valuable animals have helped countless Floridians adapt to sweeping changes in political and economic landscapes. High unemployment and low property values following the housing market crash led many entrepreneurs to breed animals and trade them through informal economic networks.[11] While interviewing these entrepreneurs in Florida and other parts of the United States, I found a large community of people who care for tropical birds, endangered frogs, venomous snakes, and lizards in their own homes. Capital investment is necessary to get breeding projects off the ground: specialized shelters, veterinary care, and food can prove costly. Providing entertainment, love, and affection for these animals also involves difficult daily labor. Flexible labor processes have enabled a diversity of people to channel their love and passion for particular species into moneymaking ventures. As a result, spectacular life forms are multiplying in garages, studios, and basements—producing emergent ecological communities within human-dominated landscapes.

Multispecies families in Florida often involve queer sensibilities, in the sense that they are based on relationships involving choice and love. Biological ties are often decentered in gay and lesbian notions of kinship, according to Kath Weston's book, *Families We Choose*. Choice in multispecies families is often asymmetrical, with humans keeping others in conditions of captivity. Still, critters that are folded into human families often choose to reciprocate by returning the love, affection, and social advances of other household members. Imagining new forms of relatedness, Janet Siskind argues, "Sociability is a panhuman, panprimate, and panmammalian tendency, but the molding of this sociability into forms of kinship and marriage is nonintrinsic and external, potentially as discardable as the chrysalis of tomorrow's butterfly."[12] As people project concepts of kinship beyond the realm of primates and mammals, some creatures share our sensibilities of sociability to a greater or lesser degree. In other words, certain species are more flexible as persons than others.

Animals not amenable to being treated like persons often die in the hands of loving humans who do not know how to care for them. Influential environmental organizations that advocate for more regulation of the wildlife trade largely focus on the relationship between capitalism and death. Under pictures of tiger rugs and elephant skulls, a World Wildlife Fund web page reports, "Illegal wildlife trade is driven by high profit margins and, in many cases, the high prices paid for rare species. Vulnerable wild animals are pushed further to the edge of extinction when nature can't replenish their stocks to keep up with the rate of human consumption."[13] Making similar critical observations, Nicole Shukin argues that global capitalist markets "fetishize animal alterity" and keep creatures in a perennially undead state of "interminable survival."[14] While some wildlife trade networks certainly generate mass death and indeed push some species toward the brink of extinction, other articulations of capital and life are producing promising futures. Committed breeders often bring ethical principles together with capitalism to animate rare forms of life—giving intergenerational gifts to species they love.[15]

Opportunistic animals, critters that have become ontological amphibians, are able to get inside human social worlds. Still, as Geoffrey Bowker and Susan Leigh Star would remind us, classifying such animals as fictive kin has consequences. Multispecies families are spaces where social persons experience torque as their modes of being are twisted, with other kinds of agents pulling at them and making demands across the species interface. When social expectations are aligned, and demands are quickly met, the torque or stress evaporates.[16] Some members of multispecies families enjoy spaces of happiness, where lively encounters bring the "the hap of what happens" into "the gap between the impressions we have of others and the impressions we make on others."[17] For others, happiness emerges only with the possibility of escaping uncomfortable familial entanglements. In Florida, surprising forms of life are running wild, as divided forces produce momentary openings into the beyond. Unruly ecological assemblages are emerging as critters escape their attachments with familial units and wander through jungly backyards to novel multispecies communities beyond human spheres of management and control.[18]

DESIGNER BABIES

One woman in her early sixties whom I met in central Florida, Caroline Elliot, helped me understand the home economics of multispecies families orbiting around tropical birds.[19] Caroline introduced me to her "babies," who live in a makeshift aviary on the back porch of her double-wide mobile home as well as in assorted cages strewn through her backyard amid a collection of salvaged lawnmowers, motorcycles, and golf carts. One of her favorite babies, Myrtle, a Hahn's macaw (*Diopsittaca nobilis*), readily engages in mimetic play.

> Caroline: Myrtle, Wanna grape?
> Myrtle: Kuluroraow.
> Caroline: No, not cold, it's warm out. Wanna sing like Nanna? Sing!
>   Ah, ah, a a aaaah! Sing, Myrtle.
> Myrtle: Kuluroraow.
> Caroline: Nah.
> Myrtle: Ah ah, a a aaaah! Wanna apple?
> Caroline: [Laughs] Give Nanna a kiss.

After kissing Myrtle, Caroline explained that birds learn to make a little noise, "mwah," when we pucker, since they carefully listen to people. "Right before they get to your mouth, they'll open their beak and just kind of push your lip," she explained. Caroline and Myrtle arguably share a social sphere, a bubble of happiness. Friends and family call her the "bird whisperer" because of her ability to work with animals that others regard as mean or intractable. Since injuring her back and being forced into early retirement, Caroline has sustained her household by buying, breeding, and selling birds. Part of Caroline's work involves scouring flea markets and Craigslist for animals that are available for below-market prices or free to a good home. Working with animals who like to bite, Caroline teaches them how to be members of human households—with appreciation for our rules about politeness, touch, regard, and response. By transforming her babies into social beings, who are attentive to human needs and desires, she adds to their value as commodities on the open market.

David Schneider insists that "kinship is social or it is nothing."[20]

Schneider assumes that kinship involves the socialization of biological processes.[21] Playing with the boundaries of kinship, mixing biological and social relations in creative ways, Caroline proved to be ever ready to draw new people and birds into her flexible family network. While we were interacting with Myrtle, she received a phone call from Jean, the former owner of Polly, also a Hahn's macaw (*Diopsittaca nobilis*). "Jean is like my sister," Caroline said. "She's interested in doing trades. I love doing trades, especially with her. She has a thing: two for one. I give her one bird—she gives me two." This phone call gave Caroline an opportunity to further describe her skills as a bird whisperer, a savvy entrepreneur, and a social networker. These two-for-one deals were on offer because Caroline was capable of making birds more social and personable.

As Caroline talked about her extended network of fictive kin, a little Australian budgie named Peeps flew out of nowhere and landed on her hand. While Caroline continued to gush about her favorite babies, I considered the serious possibility that she was simply making anthropocentric projections by addressing nonhuman animals as kin. Research on kinship by anthropologists who study people have historically projected ethnocentric, Eurocentric assumptions onto the social life of others in Asia, Africa, and the Americas.[22] As Peeps and Caroline playfully interacted— as he lightly pecked at her fingernails and rapidly rubbed his tail feathers over the knuckle of her thumb—I suddenly realized that this small bird was projecting norms about the sociability and sexuality of his species onto a human. When Peeps finished and fluttered off, he left a small droplet of semen, with a bright red dot of blood, on Caroline's hand.

Wiping her hands on a nearby towel, letting this kinky moment pass without comment, Caroline did not miss a beat and continued to tell me about her relationships with different members of her household. "I don't think that I'll ever try to breed Myrtle," Caroline said. "How come?" I asked. "'Cause she is a pet." As Caroline continued to talk, I came to understand that birds are not endlessly flexible as persons. Birds can become human persons in a way that limits their capacity for social relations with their avian kin. Later I found a website that explained: "Many Hahn's Macaw breeders will allow their parrot pairs to raise a few of their young as they make wonderful breeders as adults. A Hahn's Macaw that has been tamed and imprinted on humans will probably never breed successfully. So for this reason the best Hahn's Macaw breeders do not allow the parrots to get too tame."[23]

Thom van Dooren has written about the ethics of imprinting—a psychological process of attachment to a parent that happens early in the life of many birds. "Imprinting is not like many other human-bird inter-actions," writes van Dooren. "It is not about the formation of a relation-ship between two subjects, who—however unequally positioned—al-ready have a significantly well-formed way of life, a way of being in the world produced through particular biosocial inheritances and individual experiences. Rather imprinting enters into and disrupts some of these modes of inheritance, taking advantage of an ontological openness to produce an altered way of life."[24] The altered way of life that Caroline shares together with multiple species of birds has certainly involved foun-dational processes of torque, with the species being of macaws, budgies, and parrots twisted into a mode of sociability that is friendly to humans. Caroline's own existence has certainly also been torqued by the multitude of birds in her care, who make daily demands on her time and money. De-spite the asymmetries structuring Caroline's multispecies family, I found able evidence of happy interactions characterized by improvisation, play, and kinky interspecies sex.[25]

Caroline also introduced me to Dale, a rubino Bourke who is one of her other friendliest babies. As Dale flew around the back porch, comfortably perching on my shoulder, Caroline revealed that some of her improvisa-tion and play involved tinkering with the genetic building blocks of life. "Dale is a good boy. But birds like him don't exist in the wild—Bourkes are normally all blue, not like this striking red and yellow like the rubino." Caroline showed me the Bourke Genetic Calculator, an online tool for do-it-yourself (DIY) geneticists, which features a photographic "Muta-tion Gallery" illustrating the effects of different dominant and recessive genes. The calculator lets users enter the genetic characteristics of the parents—like the recessive genes *cin* for cinnamon, or *fd* for faded, as well as dominant genes like *V* for violet, or *D* for dark. After entering the genetic information for the breeding pair, you hit the Calculate button for a prediction of what your designer baby will look like. Caroline showed me the pedigrees, which looked like human family trees, that she uses to keep track of these dominant and recessive genes. Following time-tested principles of Mendelian genetics, Caroline knows that if she seeks to have babies with a certain trait—like *cin* for cinnamon—she needs to breed two adults who both carry the recessive version of the gene.

The DIY ethos of animal breeding has resulted in the creation of make-shift family laboratories where age-old Mendelian principles of genetics are coming together with novel technologies for making sense of DNA. Contemporary animal breeders, like Caroline, use practices that grew out of nineteenth-century European methods for calculating fitness, documenting reproductive performance, and marketing genetic capital with financial instruments.[26] Animals became objects of curiosity in Victorian England, according to Harriet Ritvo, as breeders and naturalists tested the limits of human abilities to manage "the chaotic and unfathomable variety of nature."[27] Lately such experiments have both intensified and veered off along diverging lines of flight. Caroline told me about some of her breeding projects that were relentlessly focused on preserving established pedigrees, or "wild type" genetic populations that she imagined to be from a realm that existed outside human agency and action. Her other homegrown experiments used strategies of hypercultivation or techniques of hybrid crossing to play with species boundaries, to create new wild forms of life.[28]

Emergent genomics technologies are starting to give people like Caroline more tools for their breeding experiments. Sending a bit of eggshell, a feather, or some blood to Exon Biotecnologia, based in São Paulo, Brazil, lets Caroline learn basic genetic information about her babies. She showed me the Portuguese-language Exon Biotecnologia envelopes, which she decoded with help from Google Translate, that allow her to select the sort of information she would like to learn about a given bird. For freshly hatched babies, she can learn about their paternity or gender. Groups of adults that she buys on the open market can be tested to see if they are closely related as siblings, cousins, parents, or offspring. Exon Biotecnologia can also determine the species identity of birds, including hybrid mixes. The tests do not let Caroline determine the presence or absence of mutations of specific genes, like fd for faded, or V for violet in Bourkes. But definitive knowledge about the relatedness of individual birds allows her to build better family trees, make better guesses about who has which genes, and make better decisions about breeding to produce desired color combinations.

Feelings of ambivalence certainly underpin some of Caroline's breeding experiments. She showed me some freshly hatched cockatiels, whose closely related parents had won prizes at bird shows. The hatchlings looked sick, even to my untrained eye, and she confirmed that their siblings had already died. "I think it is a problem because their parents are too closely related, so I am separating them," Caroline said. "I am generally against breeding relatives, but everyone with show birds promotes inbreeding, saying, 'Oh no, you can't bring in outside birds.' But after what is going on with these two, I don't think that I'm gonna breed relatives ever again. How would you feel if every time you had a baby it died?" At the same time that Caroline appeals to claims of authority with direct

FIGURE 8.1. Caroline Elliot produced inbred cockatiels, offspring of prize-winning show birds, that were visibly sick. Photograph by Eben Kirksey.

access to the science of genetics, some of her breeding experiments are also undermining the authority of taxonomic scientists.[29] Caroline told me about her intention to cross a nanday conure with a sun conure to produce a hybrid, a creature called a Sunday conure in the pet trade (*Aratinga nenday* × *Aratinga solstitialis*).

While some people in online avian forums defend the sanctity of species boundaries, colorful hybrids continue to fetch top dollar in a speculative economy that values novel and rare forms. Caroline takes pleasure in the confusion of some species boundaries, while also making arguments for responsibly maintaining other distinctions.[30] Ethical commitments centered on the welfare of individual animals, rather than an absolute defense of all predetermined species boundaries, guide Caroline as she pairs her babies together. Caroline's story of rescuing April, a hybrid pairing of a sun conure and a peach-fronted conure (*Aratinga solstitialis* × *Aratinga aurea*), illustrates the embodied problems that emerge when certain species boundaries are irresponsibly crossed. "The guy who gave me April said that the parents pushed her out of the nest and that messed up her hip. But I found out through investigation on Google that almost all of the sun–peach-fronted hybrids come out with that same hip deformity.

It's a genetic thing." Caroline would never try to repeat such a mistake. She would never deliberately create a bird with known genetic problems.

As I experienced the full tour of Caroline's double-wide mobile home, I noted that some species of birds—the macaws, African greys, Bourkes, and conures—were more flexible, as persons, than others. Some individual birds, like Myrtle with her ready ability to mimic human speech and her social responsiveness, were more personable than others. To play with my own queer jargon, these animals were social amphibians, able to intra-act with other beings in multiple worlds. By comparison, her Gouldian finches, diamond doves, parakeets, and cockatiels were less sociable with people, less attentive to Caroline's desires for touch, interaction, response, and regard. Caroline was thus attached to some of her babies more than others. Even still, in tough economic times, she could not always ignore the exchange value of some of her most treasured loved ones. She cried when she told me about selling Randy, a $5,000 talking Catalina macaw, to pay an overdue medical bill.

Many other bird enthusiasts I interviewed in Florida shared similar sentiments—they were attached to the animals who were part of their families. Their own feelings of well-being had become contingent on the well-being of the creatures in their care. When contingencies beyond their control resulted in the death of a loved bird, people experienced intense emotional distress.[31] Some pet owners who were not involved in breeding expressed horror at the thought of selling their companions on the open market. Still, everyone I talked with understood that the animals in their families were vulnerable to sudden shifts among humans in the household. With a death, a divorce, a lost job, or a child headed away for college, the status of prized birds could suddenly be demoted—presenting an opportunity for someone like Caroline who was always on the lookout for valuable orphans.

Sometimes sudden shifts in human households gave the animals themselves the opportunity for freedom. One person told me, on the condition of strict anonymity and confidentiality, of how her enraged husband opened up all of the bird cages one Saturday while she was out at the beach with friends. The husband had grown tired of the incessant squawks and screeches and could not convince her to sell the birds. They have since divorced, and now she has assembled a new multispecies family as a single mom with both birds and human kids. She takes pleasure in imagining that her loved ones are now flourishing in the wilds of Florida. Smiling, she pulled her *Peterson Field Guide* off the shelf, and pointed to the page for parrots and parakeets—family Psittacidae. The only member of this family that was native to North America, the Carolina parakeet (*Conuropsis carolinensis*), is now extinct—last seen in Florida in 1920. A diversity of others have taken its place. By the latest count, there are seventy-two members of the Psittacidae family in Florida: endless varieties of lorikeets, lories, galahs, budgies, cockatoos, macaws, and lovebirds, in addition to parrots and parakeets.[32]

My anonymous source told me that she is amused that her own broken family has created the possibility for new hybrid families. Her husband released her golden conures (*Aratinga solstitialis*), an endangered bird from Brazil, which may have an ability to hybridize with closely related species already wild in Florida: the Hispaniolan parakeet (*Aratinga chloroptera*, Hispaniola), and the orange-fronted parakeet (*Aratinga canicularis*, Mexico). Some designer babies, rosy Bourke mutants (*Neopsephotus bourkii*, Australia) and a lutino cockatiel (*Nymphicus hollandicus*, Australia), were also released by her husband. Florida has become home to diverse animals that were once separated by geography and history. Novel kinds of creatures are emerging within these contact zones, giving rise to new symbiotic associations.[33] Even as some state officials still talk about culling wild rhesus monkey populations, there have been no recent campaigns by the state to eliminate these nonnative birds. Multispecies foraging flocks comprising different kinds of parrots and parakeets are now a common sight in Florida. The fickle play of human desires and capital has produced emergent ecological communities.

SPECTACULAR COMMODITIES

While some birds were flexibly incorporated into human families, like the dogs Shir-Vertesh describes in Israel, other kinds of critters in Florida were resisting attempts to make them into persons. As I ventured into the realm of reptiles, I heard stories about household tensions caused by the presence of snakes. Zeb Carruthers, a snake enthusiast who got his start with pit vipers, told me that his wife forced him to make a difficult choice when his son was born: either get rid of the venomous snakes, or lose her and the baby.[34] Certain animals, his wife vehemently maintained, are not good for people to live with.[35] After selling his pit vipers, Zeb gradually convinced his wife to accept another species of snake in the house: *Chondropython viridis*, or chondros for short.[36] These new snakes also had problems assimilating to his family. This time the problem was the snakes, rather than Zeb's wife. The chondros often became agitated every time he went in their room, sometimes striking out, hitting the glass of their enclosures. These snakes also defied Zeb in the only real way open to them: they lost their will to live. After they gradually stopped eating, Zeb's three chondros—worth $300 apiece—died, one right after the other.

Chondros are known to be persnickety eaters. A book by Greg Maxwell, *The Complete Chondro Python*, describes a technique for getting these snakes to eat, which he euphemistically calls "assist feeding." This involves taking a dead baby mouse or rat (called a pinkie), moistening it with water, and forcing it down the snake's throat. The snake is grabbed by the back of the head, then the pinkie is pushed into its mouth with a pair of blunt-tipped steel forceps. "Carefully push [the pinkie] past the rear jaw line and into the throat," Greg Maxwell advises. If the snake spits the pinkie out, then the process begins again. "I have a rule when I assist-feed: 'Greg always wins.' Babies hate being assist fed and will resist vigorously." Making these unhappy animals live, Greg Maxwell has learned to apply Michel Foucault's core principles of biopolitics—about using power over life to deny the right of death.[37] "Capturing, restraining, and forcing food down their gullets does expose them to risk of damage," Greg continues, but "I assist-feed to save them."[38] These acts of supposed salvation are also performed by Greg Maxwell because chondros are very valuable on the open market.

At the Daytona Beach National Reptile Breeders' Expo, one of the largest and oldest gatherings of its kind, hosted annually in Florida's famed Spring Break vacation destination, I found a dedicated community of chondro enthusiasts intermingling with breeders of bigger snakes like ball pythons and boa constrictors. At the expo I heard about the legendary Trooper Walsh, who used inbreeding to produce one of the very first strains of designer chondros, the blue line, in the 1980s. Walsh's snakes caused a sensation. One sold for $100,000. Chondros were new on the market in the 1980s, freshly imported from West Papua, a territory under Indonesian military occupation where few foreigners were allowed to venture— then and now.[39] By 2011, when I visited the Daytona Beach Expo, the price of chondros had come down out of the clouds. But they were still fetching top dollar. Sam Reynolds, a gregarious clean-cut fifteen-year-old chondro enthusiast who flew in for the event from New York City, told me that he was attracted to these animals because "they are really expensive, they are hard to find, hard to breed, and have the coolest color patterns of any snake in the world." In total, Sam has already invested $10,000 in purchasing snakes—not to mention the money he spends feeding, housing, and caring for his animals.

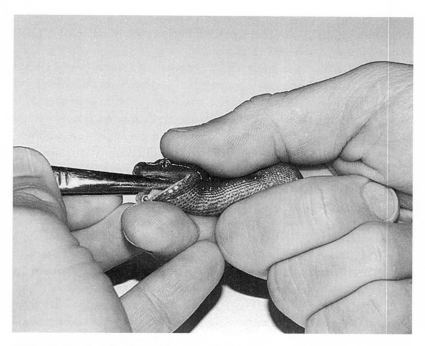

FIGURE 8.2. People who keep *Chondropythons* often force-feed their snakes to keep them alive. The original caption of this image, which is reprinted from *The More Complete Chondro* by Greg Maxwell, states: "It is important to get the pinky as far down the throat as you safely can when assist-feeding a neonate. This reduces the chances of the baby spitting it back out. Grip the body with your fingers to keep the neck straight and the rest of the snake under control."

Sam has carefully studied what he calls "the gamble of ontogenic color change" in baby chondros. The underdetermined ontogeny, or development, of this species has led Sam, and countless other enthusiasts, to place money on buying snakes that might generate colorful surprises. Baby *Chondropythons* are generally more colorful than fully developed adults. The young snakes, or neonates, are either dark red or bright yellow, with distinct black-and-white stripes. "Biaks generally throw out reds," he said, referring to a color morph named after Biak Island, a place where I witnessed a massacre in 1998.[40] While Sam had little knowledge of West Papua's history or geography, this smart teenager impressed me with his knowledge of the speculative economy surrounding this snake species: "A dark stripe on a neonate generally predicts that it will become blue. Buying up neonates is a gamble that can really pay out, or really go south." Going south, in this case, means turning green. Another common name for chondros is the green tree python, but the normal color of these animals is not valued by enthusiasts.

The ontogenic indeterminacy of chondro colors was part of what prompted Sam's speculative investments. The promise of unexpected becomings in future generations of his snakes must also be figured into his gambles. The eight chondros in Sam's collection contain a riotous assemblage of color genes with material from most of the major designer lines. In addition to snakes whose lineage he traces to Trooper Walsh's blue line, his animals have genes from Greg Maxwell's Calico Project (established 1993), Damon Salceis's albino python (produced in 2002), and the Lemontree Project, a venture whose tangled genealogical roots are lost in the mists of chondro mythos. Through the pedigrees of these snakes, Sam has established his own ties with a distinguished and distributed kinship network.[41] Mixing and mating snakes from distinct lines, Sam hopes to make his investment back with a single clutch of eggs. Chondros can lay up to about twenty eggs. Each neonate from designer lines can easily fetch over $1,000. Rather than valuing animals who might be good to live with, flexible creatures like the birds of Caroline Elliot who can become social persons, Sam talked excitedly about snakes as commodities—valued only for the price they might command at the Daytona Expo, in an online marketplace, or for use as breeding stock. Chasing after elusive genes that encode for pixilated color breaks and dreamy calico patterns, Sam hoped to produce ever more spectacular animals.

Spectacles are like commodities, according to Guy Debord, since they involve asymmetrical social relations that produce feelings of alienation.[42] Debord suggests that "the spectacle is not a collection of images; rather, it is a social relationship between people that is mediated by images."[43] Spectacles, in strictly human realms, alienate spectators from contemplated objects when illusions are staged that promote false consciousness about social life. Exposing the dynamics of these illusions, Debord worked to describe how spectacles were created by the "oldest of all social divisions of labor, the specialization of power."[44] As I navigated social relationships among people that were mediated by snakes, I concluded that the spectators were united by the objects they contemplated and created. They had formed an emergent biosocial community, a bubble of happiness and speculation, around their spectacular objects.[45] But the objects themselves, the snakes, were certainly alienated from the excited and happy affects at play in Daytona's Convention Center. The reptiles were out of sync with the happy feelings of the humans.[46]

The bubble of financial speculation at the 2011 Daytona Expo also included ball pythons (*Python regius*), a distant relative of chondros from Africa. The name "ball python" refers to the animal's tendency to curl up into a ball when stressed or frightened.[47] One ball python breeder, whom I asked about the pricing rationale for his $65,000 snake, told me, "It's as complicated as the stock market." This economic system, in the words of another high-end ball python breeder, was all about "fun and anticipation of new color morphs." Ball python morphs have names that evoke their colors and patterns: I encountered bananas, pinstripes, deserts, spiders, spinners, mystics, and phantoms. In contrast to the careful ethics of Caroline Elliot, who was attentive to the welfare of her hybrid and mutant avian babies, I found few ethical voices amid the wild play of capitalism on the bodies of frightened snakes.

Most of the chondros at the Daytona show were curled up on their perches. These nocturnal animals were apparently trying to sleep and ignore the chaos as hundreds of noisy humans crowded around them. One established chondro breeder at the Daytona Expo, Anne Alison from Minnesota, told me that she is critical of her friends and colleagues who place making money above the welfare of their animals. "These breeders will do anything to keep high-end designer snakes alive," Anne told me. They like to "pip" their precious eggs, to cut open the shells before the snakes hatch. "I refuse to do this," said Anne. "If a snake is born without an egg tooth, and we help it survive, then we are introducing things into the pet trade that can't survive on their own." Anne told me that the designer snakes themselves were vulnerable. These life forms were created by humans and have come to depend upon us for their very survival.[48] People who inbred their snakes, or promoted genes associated with health problems, were creating sick animals that would suffer their whole lifetime, she said.

Rather than creating inbred animals with colorful patterns, Anne was envisioning future generations of chondros that would feel comfortable living in captivity. Corn snakes once fetched high prices because they were hard to breed, she said. This species became a common pet. Breeders created lines of corn snakes that flourished in human-built life support technologies. Green tree pythons will never be corn snakes—Anne has lost more chondros than any other snake. "No snake in the world is used to seeing these big primates hovering over them all the time," Anne said. "They don't like to live in tiny see-through rectangular boxes, to eat thawed rats." But by breeding only healthy chondros, she was attempting to increase the viability of this species in the domestic ecosystems that have been created in human homes. In contrast to many birds, mam-

FIGURES 8.3 AND 8.4.
A $5,500 ball python exhibiting the typical fear response for this species. Chondros, arboreal snakes from West Papua, form their own characteristic balls around branches. Photographs from the 2011 Daytona Beach National Reptile Breeders' Expo by Eben Kirksey.

If it turns green....I'll replace it!

mals, or insects, snakes do not generally have prolonged social relationships. "Snakes are usually solitary creatures and are not comforted by
interaction with their owners or even other snakes," according to one
popular website, acreptiles.com. "When they encounter another animal in
the wild, they're either thinking about eating it or being eaten by it—or
breeding it. :-)"[49]

As Anne Alison dreamed of a better future for chondros in captivity, I
met others whose snakes had already learned to live more or less happily
alongside humans. Wayne Smith, a plumber who had tattoos down the
length of his arms, showed me his snake and gave me a window into his
queer multispecies family. Wayne's chondro, a light green animal with
subtle yellow hues and speckled blue dots called "mites," was out of its
cage, gently and comfortably curled around his arm. Wayne took relish in
telling me a story about sex and snakes. He recounted calling up his boss
one day to say he wouldn't be coming in to work. A pair of his chondros
had been locked together for several days while breeding—the larger female was dragging a smaller male around the cage, who simply refused to
let go of her cloaca with his penis. By the time Wayne was able to separate
the pair, the male's penis was inflamed, in need of urgent veterinary attention from a specialist doctor who lived out of town. "Let me get this
straight," Wayne mimed his boss in a thick Southern accent. "Are you
telling me that you can't come to work today because your green tree vi-
per has an infected penis?" Wayne's boss let him miss work and probably
enjoyed retelling the story to his own friends. "Most of my neighbors in
northern Florida think that I am a pee-drinking, Satan-worshiping freak,"
Wayne added.

Under a veneer of bravado, Wayne revealed deep feelings and affection for his snake. Clearly he was skilled at caring for chondros and was
attentive to the needs of the animals living with him in a common household. Like Caroline Elliot's birds, part of what made the snake valuable
to Wayne was its ability to respond to human desires for touch and contact. Care and love were central to Wayne's relationship with his snake,
even though he was not sure that these feelings were reciprocated. Wayne
Smith was not convinced that his snake recognized him as a person, a
social being, or even a living creature. This uncertainty about his own
status in the snake's eyes was part of what led Wayne to give it up for
sale. Wayne also told me about his plans for an epic motorcycle journey
with friends, something that would simply not be possible with a snake.
Following the logic of Haraway's lively capital, the capacity of Wayne's
snake to tolerate human touch, its encounter value, should have increased

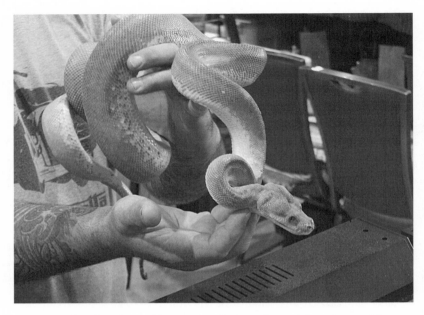

FIGURE 8.5. The chondros of Wayne Smith (pseudonym) have learned how to interact with him, to respond to human desires for touch. Photograph by Eben Kirksey.

its exchange value. But a small sign at his Daytona Reptile Expo stand indicated that he was having trouble making a deal. The sign read, "Sale, Sale, Sale. Everything must go!!! Any reasonable offer will be considered." Instead of lively capital—with commerce and consciousness, ethics and utility all at play—at Daytona I found that snakes were simply valued for novel color patterns on their skin and their underlying genetic code.[50] Capital exhibited a fickle spirit at this trade expo—possessing, and then abandoning, the bodies of vulnerable animals.

Multispecies families are constantly breaking apart amid shifting market forces and changing personal circumstances. State agents are standing by, ever ready to help. Florida's Fish and Wildlife Conservation Commission has a few dedicated employees whose core work involves transforming one household's unhappy disaster into a happy opportunity for others. At the Daytona Expo I met up with Jenny Novak, Florida's exotic species special projects leader, who was responsible for fostering continuity of care for nonnative animals amid shifting family circumstances. Jenny's job involved speaking with pet owners like Wayne on a daily basis and working to help place animals in the hands of responsible caregivers. She periodically hosted an event called Exotic Pet Amnesty Day. "These events give people with unlicensed animals, even something like a tiger acquired on the black market, the opportunity to surrender their animals with no questions asked," she told me. With a blanket policy that gives amnesty to illegal alien animals, the aim was to reduce the number of orphaned snakes abandoned by the side of the road, or birds turned loose on the wing.

When I asked Jenny about the seventy-odd species of parrots living in Florida, as well as reports of established vervet monkeys (*Chlorocebus aethiops*) and squirrel monkeys (*Saimiri sciureus*) near Miami, she pointed me to a classic paper in the journal *Ecology* that defines the principles of the Tens Rule. In evaluating the success of animal and plant "invaders," this paper found "the statistical rule holds that 1 in 10 of those imported appear in the wild (introduced or casual), 1 in 10 of those introduced become established, and that 1 in 10 of those established become a pest."[51] In other words, according to Jenny, "It simply isn't a 'sky is falling' situation." As for reptiles, she told me that Florida was once home to just two species of anole lizard. In Cuba there are fifty kinds of anoles, which each inhabit different niches. "Now the brown and the green anoles are just using different parts of the environment, living alongside other introduced species," Jenny said.

Decisions to embark on an eradication campaign are only made by the state of Florida after carefully evaluating evidence that a given species poses a risk to agriculture, human health, or native wildlife. "We only do eradication when it is feasible and necessary," Jenny said. She told me about how situated political, economic, epidemiological, and ecological concerns came together in the decision to eradicate the Gambian pouched rat (*Cricetomys gambianus*), which was linked to a monkey pox outbreak in 2003. This was a rare story of a successful eradication program, largely because the Gambian pouched rat populations were confined to a small island.[52] With other species that have escaped into the wilds of Florida, like the infamous Burmese python (*Python molurus bivittatus*), decisions to embark on population management programs have been made amid much media fanfare, even if they have little chance of achieving the goal of eradication.

Shortly after I saw Jenny Novak at the Daytona Reptile Breeders Expo, she began heading the Python Patrol, a new program of the Florida Fish and Wildlife Conservation Commission. She began traveling throughout south Florida to train anyone who worked outside—from maintenance workers in parks to employees with Florida Power and Light—to recognize and report Burmese pythons. Jenny also began teaching members of the public how to catch live pythons. Working with people who sought out dangerous and risky encounters with animals, Jenny helped bring quietness to wild situations. Generally she discouraged people from killing snakes.[53] Whenever possible, Jenny tried to twist the imminent possibility of death for individual animals back toward life. While Burmese pythons can no longer legally live as pets in multispecies families, when situations permitted she tried to place them with licensed reptile dealers, public exhibitors, and researchers. This broader community of nature lovers and animal advocates worked with her to experiment with novel approaches to the care of Florida's damaged country amid long legacies of migration, irreversible historical accidents, and ongoing ecological changes.[54]

LETTING GO

While designer snakes worth tens of thousands of dollars grabbed all of the headlines at the Daytona Reptile Expo, there were also many frog enthusiasts on the sidelines of the event—teachers, children, self-described computer geeks. A divide separated the social worlds of people who were raising frogs and those who were breeding snakes. Prized frogs, on offer for $40 to $75, were being sold along with high-end vivariums—living ecosystems, like the one we installed inside *The Utopia for the Golden Frog*, designed with the well-being of amphibians in mind. Following frogs to people's homes, I found dedicated caretakers who were attentive to the needs of their pets and curious about their experiential worlds. Trolling through classified ads on Craigslist and other online forums, I came across websites like www.frogforum.net, with "great advice, friendly people, and lots of frogs!" This community-run website allows enthusiasts to buy, sell, and trade animals. If a biosocial world has emerged around snakes, where raw commercial exploitation of spectacular commodities is the norm, the community fostered by Frog Forum is animated by livelier forms of capital. Self-taught experts trade tips about animal care, aquarium construction, and veterinary concerns. The posts on Frog Forum reveal a biosocial community where the genetic integrity of populations, the welfare of individual animals, and other ethical issues are central concerns.

Some people who love frogs share intimate moments from their multispecies families with virtual strangers in online forums. "My new Red Eyed Treefrog took this cricket moments after I snapped this pic!" wrote a self-described "Surrogate Frog Mom" from central Florida. "Does anyone else get super excited when they see their new babies thriving and doing well? I felt like my kid was goin to his first prom, when I watched him jump into action and eat the cricket. LOL!" These feelings of happiness were difficult for me to understand until I adopted Steve, an Australian green tree frog (*Litoria caerulea*), into my own family. When Steve first arrived in the apartment I shared with my partner, I was anxious as we learned how to meet his basic needs for a secure terrarium, dechlorinated water, heat, and live insect food. Steve certainly also experienced anxiety during his first weeks with us—repeatedly hopping against the glass until he hurt his nose. More experienced frog enthusiasts taught us, through online forums and telephone consultations, how to attend to Steve's needs. As we adjusted the conditions of his enclosure, his wild escape attempts settled into quieter comportment.

As our attachment to Steve grew, we found pleasure as we learned to support his well-being and flourishing.[55] Consulting with studied hobbyists on Frog Forum and other experts, we worked to bring elements of happiness—as in good hap, good fortune, or happenstance opportunities—into Steve's life.[56] Jodi Rowley, an amphibian biologist at the Australian Museum, told me that green tree frogs often sneak into human houses in northern Australia, taking up residence in toilet tanks and water cisterns. As long as no pesticide or toxic cleaner residue is lingering in these domestic spaces, our homes are generally good habitat for this kind of frog. We occasionally let Steve roam about. He showed a predilection for being near my partner rather than me.[57] When she talked on the phone, Steve would often sing along—"raaaan, raaaan, raaan."

For frogs to live happily in human households, caretakers must learn species-specific norms of tact, politeness, and what Matei Candea calls interpatience. Rather than intimate interactions based on constant touch—as one might expect from a dog, a cat, or even a bird—companion animals like frogs demand a certain distance. Being patient in interspecies relationships with organisms who are less inclined to be personable involves rethinking the tempo of "being with" and "being together." Achieving relationships of interpatience in a multispecies family with frogs also involves a recognition that predation is the primary social interaction in amphibian worlds.[58] Green tree frogs are exceptional among amphibian species in their ability to tolerate occasional human handling. As long as your hands are wet, and free of any soap or chemical residue, these interactions can be safe for all involved. When Steve was in an interactive mood, and we showed the proper patience, he would crawl onto an offered hand. Part of our routine of care involved regularly picking him up. Steve arrived in our house with a skin infection, and we were instructed by a husbandry expert to give him several warm baths, with a diluted betadine solution, each week. It was easy to tell when he was unhappy with our handling, as he would "squirt pee," sending a jet of urine our way. Happiness was less easy to discern. When handled with careful tact, Steve seemed to enjoy his warm baths—or at least be patient in the shallow dish of water, waiting for us to take him out.

Steve was a rescued frog, an orphan who would have had difficulty surviving in an unfamiliar environment without help from human surrogates.[59] I obtained him from the FATS (Frog and Tadpole Study Group) Frog Rescue Service—a group that has found homes for thousands of frogs that arrived in Sydney, Australia, in boxes of bananas and other produce. The open market for frogs as pets is one of the forces contributing to amphibian declines. Fifty frog species are in rapid decline as a result of heavy extraction for sale as pets, according to a 2004 article published in *Science*.[60] Studying the market for rare and exotic frogs, I visited some of the premier frog retail establishments in the United States—Fauna on the Upper West Side of Manhattan as well as the East Bay Vivarium in Northern California—to interview key players in the industry. Owen Maercks, who manages and owns the Vivarium, started off by bluntly saying, "Many people in the pet trade are scumbags." Pointing me to a popular work of investigative journalism that exposed endangered species smuggling rings, *The Lizard King* by Bryan Christy, he said, "I know all the people in this book. Don't think less of me. My main goal is to breed reptiles and frogs and sell them as pets."[61] But Maercks also echoed the rhetoric of Kevin Zipple's Amphibian Ark: "When a forest is about to be turned into a farm or logged, we have locals running ahead of bulldozers, culling everything out of acreage. Catching stuff in the wild can be an act of salvation."

Departing from this claim by Owen Maercks, I began to imagine how the proliferation of multispecies families might help save the day in an era of mass extinction. Thousands of frog species are living in precarious situations—beyond the carrying capacity of cash-strapped zoos and conservation organizations like the Amphibian Ark. Perhaps the best way to "save" endangered frogs, to ensure that they have a future, would be to let a multitude of people care for them in their own homes. Novel niches have emerged in landscapes that have been transformed by humans. Endangered frogs have the opportunity to invade and occupy bubbles of comfort created by us; to generate lively futures for themselves in air-conditioned living rooms, basements, and bedrooms throughout the industrialized world. A pet craze focused on endangered frogs might have the potential to generate happy futures for legions of individual animals, while also generating capital for people who breed organisms to feed market demands.

Pet frogs certainly have curtailed freedoms within human households—they have a restricted ability to form autonomous social and ecological networks, and they have lost their capacity to flee.[62] But if happiness means living with the contingency of the world, then animals kept as pets are arguably better off than the golden frogs of Panama, the endangered species "owned" by the Maryland Zoo whose collective future has been circumscribed by the zoological community.[63] In my own home I worked to create a happy life for Steve, at least in the sense of a life with "good hap" or fortune, with occasional opportunities to explore the world beyond his vivarium. Yet my partner and I worried that Steve was being denied some fundamental elements of happiness, as he was isolated from other members of his own kind. Despite many problems associated with keeping animals in captivity, hobbyists are usually able to spend more time caring for individual frogs, as well as more money, compared with cash-strapped zoos and aquariums.[64] I was certainly better able to take care of Steve than the hundreds of endangered animals competing for my attention when I volunteered as a keeper at accredited zoological facilities.

Tree Walkers International, a group of committed hobbyists, is working with the community of people who love frogs to link ethical and legal principles to the flow of capital generated by their collective desires for pets. The group's name comes from the Greek origins of the poison dart frog genus *Dendrobates* (δένδρο = tree, βάτες = walk). Ron Skylstad, the director of Tree Walkers, told me that the pet trade has historically been one of the problems contributing to amphibian declines. International networks that just treat animals as spectacular commodities without regard to sustainability, he said, can act "like a vacuum, sucking up animals from the wild." Thwarting this vacuum cleaner effect, Ron's group has created a network of hobbyists in order to help supply more captive-bred frogs in the United States, thereby alleviating the pressure on wild populations elsewhere in the world. Recognizing that captive breeding programs in the United States can threaten the economic livelihoods of poor people in rural areas of other countries, Tree Walkers also supports sustainable harvest as part of the stewardship for some amphibian species. Generating money from the sale of wild-caught frogs, the group argues, can produce incentives to protect frog habitat.

FIGURE 8.6. In 1932, several hundred poison dart frogs (*Dendrobates auratus*) from the island of Taboga in Panama were released in the upper Manoa Valley of Oahu, Hawaii. This deliberate introduction was an attempt to control nonnative insects. No frogs are native to Hawaii, but there are currently six other frog species that have become naturalized. Photograph of a captive *Dendrobates auratus* by Hans Hillewaert in the Antwerp Zoo.

The Amphibian Steward Network, the flagship conservation program of Tree Walkers, manages self-sustaining populations of endangered frogs in the United States. Ron himself oversees breeding projects involving one species in particular, *Phyllobates vittatus*, whose distribution straddles the border of Panama and Costa Rica. In contrast to some of the more charismatic and colorful frogs, this species is predominantly black with subtle red or yellow "racing stripes" down the sides. *Phyllobates vittatus* is flourishing in the pet trade. In addition to the animals actively managed by the Amphibian Steward Network, this endangered species is also bred and sold by major private retailers, like Josh's Frogs.[65] *Phyllobates vittatus* was formally designated as endangered by the International Union for Conservation of Nature (IUCN) since "its population is considered to be severely fragmented, and there is continuing decline in the extent and quality of its habitat in Costa Rica."[66] Additional imports were prohibited (without special permits) after *Phyllobates vittatus* was added to Appendix 2 of CITES, the Convention on International Trade in Endangered Species. Despite restrictions on international trade, federal law still permits the buying and selling of this species within the fifty United States.[67]

Happy futures for some endangered species are thus already emerging in pet trade networks with management and oversight from Tree Walkers. In addition to *Phyllobates vittatus*, their Amphibian Steward Network also encompasses *Mantella aurantiaca*, a critically endangered frog from Madagascar, as well as two kinds of *Ranitomeya*, multicolored poison dart frogs from Peru.[68] While contingencies can produce happy opportunities, frogs are just like other pets that can become vulnerable amid major life events like divorces, the birth of a human child, or changes in employment.[69] Tree Walkers International recognizes this problem. "The distribution of specimens in the private sector is controlled by market forces rather than conservation strategy," writes Brent Brock, the programs director of Tree Walkers International.[70] After starting up an e-mail dialogue with Brent and other members of the Tree Walkers executive committee, I asked if frogs were ever treated as "flexible persons"—loved and incorporated into human lives, but abandoned as personal circumstances shifted. After others chimed in on the e-mail chain, Brent wrote a thoughtful note in response:

> We know there is a small army of people out there with husbandry skills that has the potential to do great things for amphibian conservation. We also know that conservation breeding is a highly specialized and technical form of husbandry. And we know that husbandry takes a commitment of time and money which can be ephemeral. When a zoo loses a staff member, they hire someone else to take their place. When a hobbyist can no longer care for their collection, they have to pare back or quit. So there are significant challenges to putting this potential army of breeders to work for conservation. They need training, oversight, and management to make sure they are meeting the technical challenges of conservation breeding. And there needs to be a system to keep captive populations stable despite the realities of hobbyists' lives that compete with their being able to provide husbandry.

Working with very limited funds, Tree Walkers International is already providing some training, oversight, and management for experienced breeders within their network. This small army of people is already helping multiple species of endangered frogs enjoy bubbles of comfort—tailor-made microcosms—throughout the United States. Rather than sacrificing individual animals using a cold genetic calculus, like the people who follow the Species Survival Plan of the golden frog of Panama, this network of hobbyists is squarely focused on animal welfare concerns in the present. Converting despair about endangered species into their own concrete hopes for a shared future, the committed members of Tree Walkers are doing the difficult practical and imaginative labor to sustain shared spaces of happiness against seemingly impossible odds. These people who care for frogs are investing their own time and money to give intergenerational gifts to species they love.[71]

Alongside the frogs actively managed by Tree Walkers International, multiple other kinds of endangered species have gone wild in the pet trade. From afar, I fell in love with one of these charismatic species: *Agalychnis annae*, the blue-sided tree frog, from the highlands of Costa Rica. Combing through posts on Frog Forum, with an eye to understanding how these rare and beautiful creatures interface with online marketplaces, I found captive-bred frogs selling for $40 to $80. Blue-sided tree frogs were formally listed as endangered by the IUCN in 2008 "because of a drastic population decline, estimated to be more than 50% since 1990."[72] These declines are believed to have been caused by the arrival of *Bd*, the deadly chytrid fungus, and egg predation by the swordtail (*Xiphophorus hellerii*), a common fish in the pet trade that has been released in Costa Rica.

International trade of blue-sided tree frogs was banned in 2010 as they were added to the CITES list of endangered species.[73] The hobbyist community in the United States consolidated the animals already in the United States before the ban and began a focused program of breeding to help keep this species alive. A 2010 article in *Leaf Litter* magazine reported that one systematic breeding project of *Agalychnis annae* had produced an astounding 2,454 eggs with a tadpole hatch rate of 86 percent.[74] But a key breeder told me that after hundreds of frogs were produced and sold, many of them died. By 2014, the population of *Agalychnis annae* had dwindled to just a few frogs distributed among three breeders. Brent Brock told me that these boom-and-bust cycles are related to the difficulties of maintaining explosive breeders. "It is very expensive and difficult to raise hundreds of tadpoles and then send many very healthy frogs out to hobbyists only to have almost all of them die," he said.

These breeding experiments suggest that some frog species can live happily in multispecies families, while others in captivity are destined for an unhappy fate. Learned behaviors, or genetic predispositions, mean that

FIGURE 8.7. The blue-sided tree frog (*Agalychnis annae*) is an endangered species that has experienced dramatic declines in Costa Rica's protected forests, like Monteverde Cloud Forest Reserve. Currently some animals still live within the pet trade of North America. This explosive breeder is available to hobbyists in boom-and-bust cycles. It also lives in polluted streams around San Jose, Costa Rica's capital. The two endangered frogs pictured here were sold on Dendroboard.com for $20 each. Photograph by Seth Kiser.

some animals are better adapted to living with humans—as some of my interlocutors in the world of chondro snakes demonstrated. In other words, certain kinds of frogs are more ontologically amphibious than others— able to exist in a stable symbiosis with humans where each species integrates "a reference to the other for their own benefit."[75] But conservationists fear that once endangered animals become too domesticated, they can never again live in the wild.[76] As the community of people who cared for blue-sided tree frogs in the United States struggled to maintain optimism in the face of the death of loved ones, news from Costa Rica made me happy and hopeful.[77] Trolling through the archives of Frog Forum, I found a DIY enthusiast who had taken practices of caring for this species into his own hands.

Leon1993, a Frog Forum user in his early twenties from San Rafael de Heredia, Costa Rica, posted pictures on Frog Forum of some blue-sided tree frog tadpoles he successfully bred at home in May 2013. In response to these pictures, another user in California wrote, "NICE TADS! However, what are you going to do with all these now that they are illegal to sell/trade/give away due to them being on the ENDANGERED LIST. . . .

I'm pretty certain its illegal in your area to even house these guys. . . . If its on the endangered list here in America we can't even house them in our local state without special permits from fish and wildlife department (usually only zoos and scientists can obtain this)." The gentleman from California, who was misinformed about how the laws applied to this endangered species, was later banned from Frog Forum for an unrelated incident. Blue-sided tree frogs are regulated by the CITES treaty, which covers international trade. But this law does not apply to buying, selling, or keeping animals within a country like the United States or Costa Rica.

In response to the inquisitive and misinformed gentleman from California, Leon1993 wrote, "I know these frogs are endangered, that's the main reason why we are breeding them in captivity, I never said I was interested in selling/trading/giving away the frogs." The gentleman in California became even more nosy: "I see you also mentioned 'WE' in this post. Who is 'WE'?" Leon1993 politely replied, "Our center is still under development but we will be opening a website full of information. I am currently not working for any conservation 'company,' this is a conservation project created by myself not a company." Independent of the CITES laws, the Costa Rican government requires that anyone keeping endangered species obtain a permit from MINAE, the Ministry of the Environment. Leon1993 had his papers from MINAE in order. After all of this badgering by the misinformed gentleman from California, other Frog Forum users praised Leon1993 for his efforts: "This is wonderful! :) Congratulations!"

I reached out to Leon1993, whose real name is Leonel León, asking for a phone interview. Leonel told me that he had stopped posting on Frog Forum, saying, "They are more interested in keeping frogs in captivity. My mentality is more oriented towards conservation." Like most people who keep frogs as pets in the United States, his DIY projects orbiting around blue-sided tree frogs involve the labor of love—an expenditure of his own time and money. Encouraging me to think beyond the realm of multispecies families, where captive frogs are kept in costly long-term holding tanks, Leonel told me about how he was making simple, tactful interventions to help semiwild frogs multiply. As a young boy he had become familiar with blue-sided tree frogs. "Once they were very common in coffee plantations near where I live," he said, "but now their population has decreased." Leonel began searching local plantations until he found the frogs again. He put a plastic tank in the dirt, under coffee plants, and waited a few days for it to fill with rain. "Some months later I visited the area at night and found four males calling above the tank." In the daylight he found five egg masses on leaves, hanging over his artificial pond. He left the tank there to help make this agricultural plantation into a livelier ecosystem.

Tactfully navigating social networks surrounding deeply entrenched political and economic interests, Leonel began caring for fragile ecological assemblages within coffee plantations. Coffee pickers began bringing Leonel frogs—especially when they seemed to be sick. After certain kinds of pesticides were applied on these plantations, he found many dead and dying blue-sided tree frogs. Plantation owners were reluctant to take any measures, like switching to new pesticides, that could negatively impact their crops. Even though he could not change the fundamental dynamics of the situation, Leonel was able to make modest architectural interventions to foster life. He began attaching small plastic tubes on coffee plants overhanging mountain streams. After leaving a few of these tubes in place to see which frogs might use them to lay their eggs, he found that they were adopted only by blue-sided tree frogs. Leonel has now attached hundreds of these plastic tubes to plants near his home, a multitude of tiny rivets, to help keep creatures he loves in the world.

Leonel also told me surprising stories about the history and current distribution of blue-sided tree frogs. "In the 1990s it was very common in the protected areas of Costa Rica," he said, "like the Monteverde Cloud Forest Reserve, and Tapanti National Park. Right now it is locally extinct from those areas, probably because of the chytrid fungus. Still, this endangered frog is very common in polluted areas. I believe that this is because the chytrid fungus is very vulnerable to pollution. They are mainly living now here, in my town—San Rafael de Heredia. They are also near polluted rivers and urbanized areas." According to the IUCN Red List, this endangered species remains abundant "in San José and suburbs near heavily polluted streams, especially in shade-grown coffee plantations and gardens. . . . Research is needed to determine whether or not this species can survive only in polluted areas, because of the ineffectiveness of the chytrid fungus in such environments. If this proves to be the case, then well-meaning conservation measures to abate water pollution could unintentionally lead to the extinction of this species."[78]

Following frogs from the United States to Costa Rica, I found multiple species enjoying ancillary benefits of worlds designed with the well-being of others in mind. As oblique powers and biopolitical regulations tried to sustain forms of life with limited possibilities of happiness, I found a multitude of critters that were finding happenstance opportunities within established agro-industrial ecosystems. Wandering from Costa Rica's highlands to the lowlands—in places where it was too hot for chytrid fungal spores to survive—I found frogs thriving in national parklands that were being shaped by divided forces. Spectacular flocks of migratory birds were emerging amid legacies of cattle capitalism, a burgeoning rice industry, and transnational schemes to produce game animals for hunters in the United States. In these blasted landscapes, plants were drawing multiple species of beings and things into lively collectives and ecological assemblages, eschewing "the metaphysical binaries of self and other, life and death, interiority and exteriority." As capitalism schizophrenically warred with itself in the hinterlands of Costa Rica, as nomadic animals ran wild, I found "assemblages of multiplicities, inherently political spaces of conviviality."[79]

CHAPTER NINE

PARASITES OF CAPITALISM

. . .

"Notebook. Check. Digital recorder, headlamp. Check, check. Shotgun microphone. Check. Thermometer. Check. Boom box. Check." A team of North American biology graduate students mount an evening expedition into a wetland in Costa Rica. They are armed with an array of prosthetic sensory devices. The objective of this foray into the swamp is to decipher the calls of fringe-toed foam frogs (*Leptodactylus melanonotus*). The males of this species have distinctive fringes on their thumbs. Females deposit their eggs in foam nests that they make at the edge of the water. During breeding season, males gather together to sing a distinctive "pip-pip-pip-pip" chorus. The research team wants to know if the foam frogs are telling the truth with their calls, or if they are bluffing. When we follow John Austin in "doing things with words," human speech often functions to convince, persuade, deter, or surprise an audience. People are often insincere—someone might say, "I declare war," when she does not intend to fight.[1] Communication systems of multiple other animal species can also contain honest, neutral, or dishonest information. Some male frogs have an ability to bluff—they trick other frogs into thinking that they are bigger, better fighters. Effective bluffers, who lower their voices, reportedly lay claim to choice territory.[2]

After recording the frog calls with their shotgun microphone, and manufacturing a CD with digitally altered, low-pitched calls, the graduate students think through semiotic theories as they trudge around the swamp with their boom box. Will foam frogs bluff by lowering their voices in response to simulated frog intruders? Mud sucks at their rubber boots and sneakers. They wade in water that reaches up over their knees. Water scorpions sting them. Leeches struggle to find a gap in their clothes. A distant pair of large glowing orbs, eye shine from crocodiles lurking nearby, reflects the light from the students' headlamps and follows their

FIGURES 9.1 AND 9.2. A fringe-toed foam frog (*Leptodactylus melanonotus*), with a close-up picture of the fringes on its front foot. Photographs courtesy of Kristiina Hurme.

movement around in the water. "This is frog one-oh-one," he says into the mic. "Frog one-oh-one. Recording frog one-oh-one. Day two of our test trials." She notes the temperature of the water and then plays a blast of digitally altered frog calls on the boom box. He stands still in the warm, dark, water with the shotgun mic pointed at frog 101.

Later, back at the lab, they download the recorded frog calls onto a laptop. While the other denizens of the research station dance merengue in the dining hall, they analyze spectrographs and run statistical tests—looking for the predicted effect. Did the frogs lower their voices in response to the digital calls? A surprise emerges from the spreadsheets: the frogs were raising, rather than lowering, the pitch of their calls in response to simulated deep-voiced intruders. Were years of research on male-male interactions in frogs wrong? Were the frogs avoiding conflict? A finding like this is the stuff that graduate careers are made of—it might result in a paper in a leading journal. But these results turn out to be an artifact of sampling. Statistical analysis reveals no significant difference. The frog calls remain unintelligible. They join the dancing. By evading capture, by speaking in a way that evaded the sense-making tools of spectrographs and statistics, perhaps these fringe-toed foam frogs enjoyed good hap, or fortune.[3] Perhaps the frogs were happier as they listened to the merengue music drift across the water, or at least less anxious. Clusters of bipedal primates had stopped wading around, tracking their movements, and occasionally snatching frogs out of the water.

This failed attempt at decoding the language of another species might also be taken as an allegory for issues that arise when humans attempt to speak for nature more generally. Bruno Latour recognizes a kinship among "spokespersons"—politicians who speak for other people and scientists who speak for nature. Calling on scholars of science and society to bring democracy into new domains, Latour asks us to explore "millions of subtle mechanisms capable of adding new voices to the chorus." He has suggested we build "speech prostheses" to "allow nonhumans to participate in the discussions of humans." Nonhumans must be helped to overcome "speech impedimenta," in Latour's mind, so that they can be seated in a new parliamentary system alongside representatives of human citizens. A surprising finding might have earned the foam frog a host of new scientific spokespeople. Animals that lie, or communicate about capitulation, are readily brought into conversations about human nature. But instead of legions of biological scientists, all the fringe-toed foam frog got was me.[4]

Following Latour, I initially tried my hand at speaking with and for fringe-toed foam frogs. But I quickly realized that few people were interested in listening. While searching for democracy in Palo Verde National Park, a wetland in the lowlands of Costa Rica where I first encountered these noisy amphibians, I found powerful spokespeople for other species. Looking in vain for an agora, an open place of assembly where all interested parties were patiently waiting their turn to speak, I waded through the swamps of Palo Verde and wandered into neighboring agricultural lands.[5] This national park was situated on *la frontera*, the borderlands, where expatriate scientists, Costa Rican farmers, and multiple species meet.[6] Architectures of apartheid—gates, fences, and infrastructures of informatics—separated neighboring communities.[7] Tracing the flight of multiple species from this frontier zone, I began to search for a central place of political assembly abroad. But instead of finding an agora, I discovered that oblique powers and surprising forms of capital were actively reworking the soil of this seemingly out-of-the-way place.[8]

ANIMAL RHIZOMES

"When we go to the tropics," writes Nancy Stepan, "perhaps as eco-tourists to see the jungle, we imagine ourselves stepping back in time, into a purer or less spoilt place than our own."[9] With Stepan's words in mind, I was surprised to learn of an environmental management regime in Palo Verde National Park that was not based on re-creating an imagined original purity, but instead was bent on preserving a specific historical moment. "Our main objective has been to restore Palo Verde to the conditions of 1978," said Dr. Eugenio Gonzalez, the former director of the Palo Verde Biological Research Station, who was once one of the most powerful spokespersons representing the nature of this picturesque landscape. There were no comprehensive studies of the ecological dynamics in the wetland in 1978, but certainly everyone remembers abundant bird life. "In the dry season there was a tremendous amount of birds, especially ducks," said David Stewart, a cattle rancher whose grandfather immigrated to Costa Rica from the United States. "The sky would just be black with them, it was incredible."[10]

Before Palo Verde was turned into a national park, hundreds of thousands of wintering migratory birds routinely visited the wetlands. Spectacular flocks of up to 100,000 blue-winged teals, 100,000 pintails, and 20,000 black-bellied whistling ducks, as well as an abundance of Muscovy ducks, white-faced whistling ducks, northern shovelers, and rose-

FIGURE 9.3. Test trials with foam frogs in the wetlands of Palo Verde National Park. Photograph by Eben Kirksey.

ate spoonbills regularly alighted in the marsh.[11] "The grass-sedge marsh was open with a scattering of palo verde trees," writes Douglas Gill, a professor of zoology at the University of Maryland. "Magenta carpets of blooming water hyacinths covered the marsh as far as the eye could see. Black-shouldered kites hovered over spots as if suspended by a wire. . . . Clouds of black-bellied whistling ducks and blue-winged teals would rise with each pass of a wintering peregrine falcon."[12]

Flocks of migratory birds had long frequented Palo Verde alongside a massive ranching scheme. For most of the twentieth century, this wetland and the surrounding forests had been heavily grazed during the dry season by huge herds of cattle—ranging from ten thousand to fifteen thousand head—as well as hundreds of pigs and horses.[13] The spectacular flocks of birds at Palo Verde were flourishing within an emergent ecosystem that had been created at the intersection of market forces as well as national and imperial political projects. When Costa Rican government officials began evicting livestock from Palo Verde in 1980, with the intention of protecting waterfowl, they were disrupting lively agricultural, ecological, and financial systems.

Cattle have been entangled with capitalism for hundreds of years. The word "cattle" originated in the mid-thirteenth century from the Anglo-French word *catel*, "property." It is etymological kin to "chattel" and "capital."[14] If Marx understood human capitalists "not as agents in their own right, but as those who personify the power of capital," then cattle might also be seen as animal familiars of this ghostly spirit.[15] Never sticking to just one environment, always seeking to roam elsewhere, cattle are nomadic companions of capital. Constantly moving among elements and promiscuously intermingling with different cultural systems, cattle embody the parasitic character of capital, enabling it to invade and occupy diverse ecosystems.[16]

As one of the first organisms imported to the New World, bovines were part of what Alfred Crosby calls the "Columbian Exchange"—the phrase he coined to refer to the large-scale transfer of biological and cultural elements between Europe and the Americas. Crosby regards cattle as agents of genocide and ecological imperialism: "The animals, preyed upon by few or no American predators, troubled by few or no American diseases, and left to feed freely upon the rich grasses and roots and wild fruits, reproduced rapidly. Their numbers burgeoned so rapidly, in fact, that doubtlessly they had much to do with the extinction of certain plants, animals, and even the Indians themselves, whose gardens they encroached upon."[17] Even as cattle enabled capital to invade lands via previously unknown exploits, this nomadic species was constantly escaping along unexpected lines of flight.[18]

Nomads are "incontrovertibly destructive or tolerant," in the words of Isabelle Stengers.[19] Capitalism, which Stengers regards as "the only truly tolerant and relativist undertaking," is ever generating new nomads—agents capable of adapting to varied political, cultural, or ecological contexts.[20] Wealthy Spanish landowners captured the productive nature of cattle and capital by grounding these flighty agents, channeling their ability to destroy and tolerate different environments. Large ranches, haciendas, were established from Alta California (present-day California and Nevada) to Argentina.[21] Following the invention of barbed wire in 1874 on the Great Plains, a technology "of violence and pain across species," Latin American cattlemen were able to fence in wandering animals, to more fully capture their world-making force.[22] The ghostly specter of capital possessed the minds of people in many different nations who called themselves capitalists, people who made the expansion of capital through exchange their principal subjective aim.[23]

Cattle became the vehicles of Yankee capitalism in Latin America. In the Guanacaste Province of Costa Rica, on the northern Pacific Slope of the country, global shifts of power in the modern world system of production had dramatic impacts on local land use.[24] The land that later became Palo Verde National Park was purchased by Wilson Stewart, a cowboy from the United States, from Costa Rican president Don Bernardo in 1923. Developments in refrigeration and transportation technologies, in the coming decades, enabled ranching families like the Stewarts to transform the forested landscapes of Costa Rica into value-added biomass that was consumed by humans in distant cities.[25]

Capitalism's promiscuous spirit began possessing the bodies of multiple species—animating their movements across the landscape only to fickly dance away.[26] The spread of cattle in Stewart's land was facilitated by jaragua (*Hyparrhenia rufa*), a grass introduced to Costa Rica from Africa in 1943 that helped destroy the tropical forest and turn it into savanna. The spiky seeds of jaragua were spread by Stewart's cowboys throughout his ranch. Jaragua became a companion species of cattle. Companion species, to dig deeper into this key phrase coined by Donna Haraway, are messmates who break bread together (the Latin root of companion is *cum panis*, with bread). They are locked in relations of use where care and killing go hand in hand.[27] Jaragua grass was not propagated because it is particularly nutritious for cattle.[28] Instead, cowboys spread the seeds of this plant because it was a companion in killing that facilitated large-scale ecological transformation. Jaragua is extremely flammable and can quickly resprout after hot fires.[29] Setting annual fires during the dry season, humans helped jaragua regenerate in their fields, cocreating an emergent ecosystem. Cowboys facilitated the spread of this world-forming grass, which moved like a rhizome into the surrounding forest.

In a botanical sense, rhizomes are the underground stems of plants that spread laterally in the soil, propagating vegetatively, sending down roots, and sending up new shoots. Strictly speaking, jaragua grass does not have rhizomes. This plant does not have underground stems. If Gilles Deleuze and Félix Guattari regard ants and rats as figural rhizomes, then perhaps jaragua also has the properties of an "animal rhizome."[30] The seeds of jaragua writhe about on the ground after falling from the plant, crawling around in all directions. This movement is accomplished by a small spike on each seed, about 22 millimeters long, that is "highly hygroscopic," meaning that it readily absorbs moisture from the air to swell up and contract in length. "A constant twisting or untwisting," writes one naturalist, "causes the disseminules to creep slowly over the surface."[31] Like cattle, this grass was constantly on the move, exploring the frontiers of new environments, seeking out new places to sprout. Capital

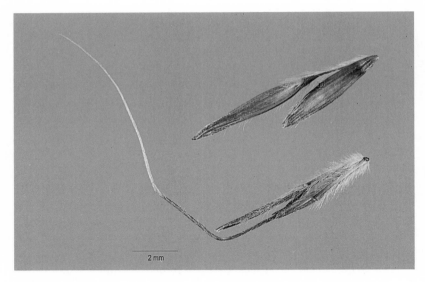

FIGURE 9.4. Seeds of jaragua grass. Photograph by the U.S. Department of Agriculture.

possessed the body of jaragua—using its talents of movement, provision, and habitat creation for cattle—to spread productive potential into hitherto impenetrable jungles. The animal rhizomes of this grass crept along the forest floor, becoming capital personified.

In the mid-twentieth century, the wetlands and woodlands of Palo Verde were firmly possessed by cattle and associated companion species. Three key players—cattle, jaragua grass, and cowboys from the United States—enfolded each other into an emergent agro-industrial ecosystem that initially seemed like it would endure for centuries to come. But as the modern world system reconfigured itself at the end of the twentieth century, and local political circumstances rapidly changed, these ecological entanglements began to unravel.[32]

Wild fluctuations in the price of beef on international markets in the 1970s began to undercut the profitability of the Stewart ranch and other cattle operations in the region.[33] At the same time, a powerful social movement of Costa Rican squatters began gaining ground in their fight for land. Making a high-stakes decision, the Costa Rican government expropriated the Stewart ranch in 1975.[34] Some of the land was divided into irrigated rice parcels and given to the squatters. The rest, some 45,492 acres, was turned into Palo Verde National Park. Social justice concerns and conservation priorities seemed to be resolved, once and for all, with the expropriation of this land. But rather than a final peace, the new stewards of the land found themselves wrangling with unexpected disruptions by multiple species of ontological amphibians. Capital began to battle with itself in the hinterlands of Costa Rica, taking on particularly schizophrenic forms.[35]

UNEXPECTED INTERRUPTIONS

After the newly acquired land for Palo Verde National Park was placed under the management of conservation biologists, all cattle were initially evicted.[36] Jaragua, the tenacious companion of cattle and capital, stayed behind in the dry uplands. Carefully patrolling the boundaries of the newly established preserve, guards worked to protect the regenerating forest and the wetland ecosystem from incursions by local farmers and hunters. After a few good years of abundant birdlife, something went wrong. Shortly after the last cattle were removed from the refuge in 1981, most of the birds took off on alternate lines of flight.[37] Other organisms also began interrupting human plans. The open grass-sedge marsh and carpets of water hyacinths came to be replaced by dense stands of cattails (*Typha domingensis*). By 1986 cattails had virtually become a monoculture in the wetlands—crowding out duck habitat, blocking views of the picturesque vista, creating a wall of vegetation that made biological research increasingly difficult. Sightings of migratory birds declined dramatically. Visions of multiple social worlds, and the lifeways of charismatic species, were disrupted by this plant, an unforeseen surprise. Lacking a spokesperson, cattails became killable.[38]

Officials at Palo Verde National Park initially imagined cattle as the enemies of conservation. But they soon began thinking about them as possible allies, inviting these nomadic animals back onto parklands once again. By reintroducing livestock, officials hoped to limit the growth of jaragua grass and cattails. If cattle and capitalism were once articulated to U.S. hegemony in Costa Rica, new political, economic, and ecological assemblages were emerging. In 1986 the national wildlife service signed a five-year contract giving a Costa Rican cattle rancher license to graze up to one thousand head of cattle in the marsh. But these cattle also proved to be difficult to manage. Unaccustomed to wetland conditions, wary of crocodiles and unknown monsters lurking in the water, the animals preferred to graze in dry grasslands where the conservationists were trying to regrow a forest. The cattle proved ineffective in killing the cattails.[39]

An early director of Palo Verde Biological Station, a biologist from the

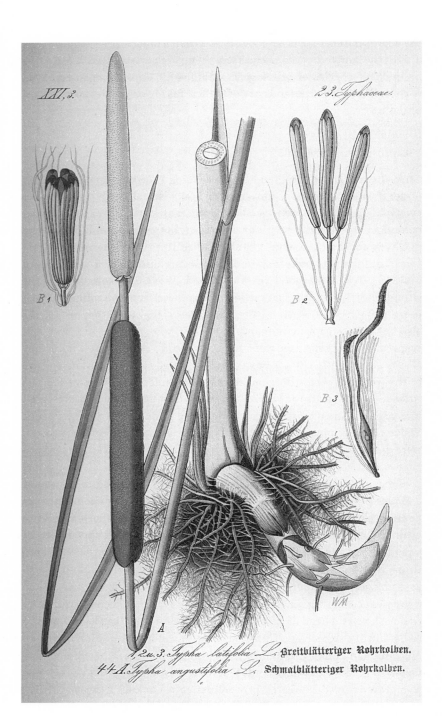

B 1

B 2

B 3

A

WM

1, 2 u. 3. Typha latifolia *L.* Breitblätteriger Rohrkolben.
4 u. A. Typha angustifolia *L.* Schmalblätteriger Rohrkolben.

United States named Michael McCoy, tried sending in machete crews to kill cattails. Rhizomes of this tenacious plant proved very difficult to combat. When chopped up, and left for dead, the rhizomes resprouted from the resulting fragments. Burning the cattails also failed. The flames destroyed nearly all of the palo verde trees, the only nesting sites for six pairs of endangered Everglade snail kites. Ashes from these fires fertilized the growth of new cattails in the next wet season.[40] Appropriating technologies of industrial agriculture, in 1989 McCoy began using retrofitted tractors with huge chopping blades as part of the ongoing effort to kill cattails and actively re-create the conditions of an earlier moment in history. This method of tilling, known as *fangueo* in Costa Rican slang, had long been used by regional farmers to cut up weeds in rice paddies, places where conventional disking or mowing techniques do not work. Fangueo proved to be highly successful at eliminating the unwanted plants in the national park. When run through dense stands of vegetation, at the right water levels, these tractors crushed the cattail stems and drowned the plants in water. In other words, these killing machines were rivets in an emergent ecosystem that helped fasten flighty waterfowl in place.

FIGURE 9.5. (*opposite*) The actual living rhizomes of cattails are more complex, fleshy, and surprising than the figural rhizomes of Deleuze and Guattari. Chemical weapons of sorts are produced by cattails. Cocktails of compounds that resemble commonly available herbicides and military-grade chemicals like Agent Orange are produced by cattails to inhibit the growth of other plant species. The characteristic brown cigar-shaped clusters at the top of each cattail spike also produce thousands of airborne seeds, enabling quick dispersion over large spatial scales. Sally Horn, of the University of Tennessee, recently found cattail pollen near the Palo Verde wetlands with a radiocarbon date of older than 2,000 years before present. In light of these findings, critical observers have begun to question the normative guidelines underlying the management regime that calls for the elimination of cattails from the national park. If the original marsh vegetation—before the cattle ranch—consisted of cattails, should this be preserved? Horn told me that evidence of cattails in Palo Verde a few thousand years ago is a snapshot in geological time. It is not a picture of the original condition of the marsh. Over geological time scales marshes come and go. The composition of plants in a marsh can change dramatically in much shorter time scales—decades, years, and, in the wake of catastrophe, days. Illustration from Otto Wilhelm Thomé's *Flora von Deutschland* (1885).

FIGURE 9.6. A fangueo tractor retrofitted with chopping blades.
Photograph by Suzanne J. Kelson.

Edgardo Aragon, a mechanic who is commonly known by his nickname, Galo, became the principal operator of the fangueo machines, the cattail-crushing tractors, in Palo Verde. Galo's trademark vehicle, an aging jeep with a bright orange paint job, is a common sight on the dirt roads that crisscross rice parcels beyond the park boundaries. Sitting on the curb outside his machine shop, occasionally interrupting our conversation to banter as neighbors passed by, Galo told me about his boyhood memories of visiting Palo Verde with his parents, family friends of the Stewarts—the Yankee cattle ranchers. "This lake wasn't always filled with a mountain of weeds, no. This lake was paradise. After all of the cows were taken out, the lake was lost," Galo said. Collaborating with conservation biologists in their war against cattails, Galo became recognized as an honorary park guard. He worked to re-create the paradise from his youth, the wetland of the 1970s.

Galo's use of agricultural techniques in this natural refuge initially produced a spectacular rebound in the populations of two species: the blue-winged teals and the black-bellied whistling ducks. In early 1991, the year tractors first plied the wetlands of Palo Verde, some twenty thousand black-bellied whistling ducks and twelve thousand blue-winged teals were

FIGURE 9.7. Galo next to his trademark jeep. Photograph by Daniela Marini.

counted in an area that had been almost devoid of bird life the previous year.[41] But Galo quickly found that the costs of operating his tractors in deep wetlands with cattails were much higher than he initially expected.[42] Gunning his motor to its maximum capacity, burning up gallons of gas and stressing his machines, was necessary to get through dense cattail stands. Manufacturing and sustaining this novel ecosystem proved to be costly. Park managers began to look abroad for funds to support their interventions with fangueo tractors.[43] Palo Verde became entangled in a transnational scheme to foster the life of migratory ducks.[44]

GAMING GLOBAL FUTURES

Ducks have powerful spokespeople. As migratory bird populations showed signs of rebounding in Palo Verde National Park, financial support for crushing cattails began to flow from Ducks Unlimited, a U.S. hunting organization that claims to be the "world's leader in wetlands and waterfowl conservation." By protecting duck habitat in Costa Rica, a place where hunting is illegal, Ducks Unlimited was generating more game for North American hunters. With more than 600,000 members, this organization is working "toward the vision of wetlands sufficient to fill the skies with waterfowl today, tomorrow and forever."[45] The Washington, DC, office of Ducks Unlimited has a team of professional lobbyists who advocate for the right to bear arms and the rights of ducks to safe and secure habitat. This organization has helped create or restore twelve million acres of wetlands.[46] With an extensive grassroots network, a well-oiled fund-raising machine, and a distinctive brand, Ducks Unlimited has brought the political into contact with what David Wesley has termed "socio-duckonomics."[47]

Socio-duckonomics refers to the commodified material culture of duck hunters, the habits of the hunted, and the value of the places where the two meet, according to Wesley. "There must be something special about four dead teals that occupy a minuscule portion of the bottom of an $8,000 mudboat," he writes.[48] Duck hunters, who place so much value in a few dead birds, harbor few illusions about the productive networks that help bring their flighty quarry into being.[49] They are also not shy about investing money in the latest technologies that facilitate the killing of these creatures. In 2006 waterfowl enthusiasts in the United States spent over $1.3 billion on guns, ammunition, travel, and recreational services.[50] The force of these funds—along with the goals of pursuing, deceiving, and killing waterfowl—has driven the development of a mind-boggling array of novel technologies: portable hay bale blinds, lead-free bullets made by the Environ-Metal company, decoys with patent-pending Breeze-Ryder bases, as well as mud boats, flat-bottomed vessels used to maneuver in hard-to-reach places in marshes and grassy lakes.

Socio-duckonomics has fueled the emergence of peculiar ecosystems throughout the Americas by optimizing and controlling populations of waterfowl.[51] In the early twentieth century, one of the major initiatives of duck hunters was to protect what they regarded as the original "duck factory"—over 300,000 square miles of wetlands in the prairie pothole region of the United States and Canada.[52] Socio-duckonomics also penetrated the U.S. federal government—becoming institutionalized in the Fish and Wildlife Service. "Through the 1980s wetlands management was centered on waterfowl," said Mark Madison, the Fish and Wildlife Service historian, in a telephone interview. "Early programs were just about counting ducks even when management practices were hurting certain ecological communities, like those of fish."

Tracing the flight of birds and the flow of capital, I landed in my own backyard. Shortly after returning from my initial field research in Costa Rica, I found myself sitting on a stuffed elephant-leg stool, at nine thirty on a Monday morning, in the office of Gary Lease, the dean of the humanities at University of California, Santa Cruz, where I was finishing my PhD. An avid hunter, a committed environmentalist, and a card-carrying member of Ducks Unlimited, he proved to be a generous and intelligent interlocutor about transnational schemes centered on fostering the life of waterfowl. Lease "knows a great deal about those he kills, how they live and die, and what threatens their kind and their resources," writes Donna Haraway, one of his colleagues in the History of Consciousness program. "His approach is resolutely tuned to ecological discourses and he seems tone deaf to the demands individual animals might make as ventriloquized in rights idioms."[53]

Running late from another meeting, Lease said, "I'm always glad to spread the gospel of Ducks Unlimited. That's what it is, of course, and like any gospel it needs a critical hearing." "Ducks Unlimited is obsessed with habitat," he continued. "Other hunting organizations, Delta Waterfowl for example, are controlling the population growth of predators—foxes, coyotes, raccoons, skunks—through trapping. Heavy trapping results in dramatic increases in the numbers of successful nests. Ducks Unlimited is only concerned with habitat modification. Some groups have recently accused Ducks Unlimited of being anti-hunter."

After chatting for more than an hour, Lease insisted that we meet the next morning at eight thirty. When I returned to his office, he was wearing a shirt with a Ducks Unlimited logo and had brought me a pile of books, magazines, and newsletters. One of the books he lent me, *Our Sport*, was graced with an anonymous poem on its back cover:

The hunters of ducks are a crazy breed,
A hole in the mud is all they need,
A place to hide from a flying duck,
In eighty acres of smelly muck . . .
Would I spend my money and waste my time,
And listen to lies in the winter time?
Would I do all these things no sane man should?
BROTHER, YOU'RE GODDAM RIGHT I WOULD!"[54]

Following a battle with cancer, Gary Lease passed away before he could tell me more about the ineffable thrill of duck hunting. Mingling with Gary's friends, who gathered to celebrate his life in 2008 at the annual California Waterfowl Banquet, I was given a privileged view of how the values of environmental consciousness and commerce at play in U.S. hunting cultures have shaped emergent ecosystems. Sixty guns were given away during a fund-raising raffle at the banquet, many in Gary's name. A general contractor, whose construction company offered up a Wesson automatic twenty-gauge shotgun to the highest bidder, told me, "Being a gun donor promotes business." Gesturing around the room, to a gambling table with Ducky Dice, to merchandise donated to the raffle and an auction, he said, "All this money is going to conservation."

As a voyeur at the Ducky Dice table, where contestants paid $20 for one roll of the dice, I struggled to understand the pleasure of this game from a safe nonparticipatory distance. The most enthusiastic player I saw was a skinny man with red ears, pale blue eyes, a muted buttoned-down shirt, and white hair. Throwing the dice with a distinctive backhand twist, his eyes sparkled and his mouth pursed in boyish glee. Winning combinations of the dice were somewhat loose. The crew running the table was free to cut deals. One young man told a contestant, "If you throw snake eyes on this next throw, I'll give you four raffle tickets." Ducky Dice participants were testing their luck twice. Raffle tickets earned participants the chance of winning one of the sixty shotguns or sundry other door prizes.

Duck hunting is a "game measured in the number of hours enjoyed," rather than just exercising skill or releasing a "bloodletting" lust, in the words of John Cartier.[55] This game is remaking global futures. It involves the contingencies of strange connections—alliances between unlikely partners, between humans and multiple other species, between enemies locked in intractable conflict.[56] Players of Ducky Dice in California, and those who enjoy the game of duck hunting, are obliquely shaping the wetlands of Palo Verde and countless other protected parklands around the globe. An uneasy articulation of U.S. hunting culture, the spirit of capital, and biology has inspired hope for some conservationists. Yet it has instilled dread for others who live in the shadows of spectacular waterfowl flocks.

LIFE IN LA FRONTERA (THE BORDERLANDS)

Following the flight of ducks led me to Bagatzí, a farming village just outside the border of Palo Verde National Park. Bagatzí was founded right after a powerful social movement, the *precaristas* (the rabble proletariat), waged a successful campaign to shut down the massive Stewart family cattle ranching operation. "I don't know how many hectares the Stewart family had," says Manuel Gonzales, an elder precarista who spent his youth agitating on the front lines against Yankee capitalism. "I can tell you in haciendas—they had seven haciendas. It is incalculable how much land they had." After the Costa Rican government expropriated the Stewart ranch in 1975 and set aside some nineteen thousand hectares of wetlands and seasonally dry forest that became Palo Verde National Park, the remaining land was divided up and given to precaristas.[57] Government institutions coordinated the selection process—soliciting recommendations from the board of the rural bank and the local political committee. "When they did the selection it was a surprise," recounted one of the lucky precaristas. "A big envelope arrived at the house that said, 'Congratulations, you were one of the people selected for the grand irrigation project.' None of our families were asking for land. It was a surprise."

Each lucky family was given a small cement house and an irrigated rice parcel when the government built this village in the early 1980s. The government also leveled ground to make a soccer field, a central village green of sorts. They built an elementary school that doubles as a church on Sundays. One enterprising Bagatzí resident opened up a bodega—a small store selling canned foods, soap, candy bars, and soda—in the living room of his house. When I took up temporary residence in Bagatzí in 2009, I was surprised to learn that the bodega possessed the only telephone in town. I also found that many of the original houses for rice farmers had been abandoned, with weeds growing on the roofs. Accommodations at the nearby Palo Verde Biological Station, where I had been learning to speak with and for the fringe-toed foam frog, were simple to be sure. Researchers were living in rooms with bunk beds, which were $65 a night, adjoining a basic laboratory and a small library with computer terminals. The biological station had only a few telephones and a painfully slow Internet connection. Still, as I moved among worlds, hitching rides with Bagatzí residents who worked at the research station, the architectures of apartheid separating these two communities were brought into sharp relief.

In Bagatzí I rented a room for $4 a day in Tifa Tours, an artisanal papermaking cooperative run by a small group of women from farming families. Scientists from Palo Verde Biological Station, and program officers from the United Nations Development Program, had been providing aid to these women in hopes of transforming the biomass of cattails (*Typha domingensis*) into value-added aesthetic objects. The donors were trying to create economic incentives for killing cattails. Ana Janzen, one of the founders of Tifa Tours, gave me a tour of the facilities in early 2009, after showing me a room full of cobwebs and bunk beds that was to be my temporary home. She told me about the heady moments when they rushed to fulfill their first big order for *Typha* paper—a request from the

FIGURE 9.8. A machine in Tifa Tours that broke down while printing diplomas for University of Costa Rica graduates. The cooperative became inactive after the machines stopped functioning. Spectral promises made to the women of Bagatzí—fictions and fabulations about modernity—evaporated as the fickle spirit of capital danced away. Photograph by Eben Kirksey.

University of Costa Rica for eight hundred diplomas for their graduating class. "Everyone worked hard to meet the deadline," Ana told me, "spending long hours here at the factory. But when we were almost done, one of the machines got really hot. And then it stopped working properly. We finished the job, but the last sheets of paper didn't turn out so well."

Cattail fiber is not particularly well suited for making paper. "We made rice paper that turned out beautifully. The banana paper was also nice. We really tried hard to work with *Typha*, because this is what our donors wanted, even though we found it didn't produce very nice paper." The papermaking machines were imagined by the donors as ecological rivets, cattail-killing machines that would help keep ducks in the world. But the schemes of the donors, who were perhaps unwitting agents of socio-duckonomics, ultimately failed. Cattails proved resistant to capture by capitalism and remained in the national park. This rhizomorphic plant broke the machines of this modest papermaking project just as the operation was starting to get off the ground.[58] Visions of turning Bagatzí

FIGURE 9.9. Duck hunting, the world-making game of powerful foreigners, is illegal in Costa Rica. Some farmers of Bagatzí have become poachers, intruding onto the government nature preserve in Palo Verde managed by conservation biologists. This mannequin with a wooden gun was installed by park guards on one of the dirt roads crisscrossing the national park. Guards made mimetic copies of themselves, aiming to startle intruders with uncanny specters. Photograph by Eben Kirksey.

into a picturesque ecotourist destination, by establishing bunk beds for visitors who might want to tour marshes full of *Typha domingensis*, also failed.

The story of Tifa Tours offers an opportunity to pause and consider the ethical and political values built into articulations of technology and biology. Ana Janzen's lived experience speaks to the opportunities and challenges that emerge when one dwells in borderlands where multiple species meet amid relations of radical asymmetry. Gloria Anzaldua's bilingual book, *Borderlands / La Frontera* describes the U.S.-Mexican border as *una herida abierta* (an open wound) where the third world grates against the first and bleeds.[59] Making paper with cattails was helping make habitat for ducks—one of the things bleeding across the frontier zone separating Bagatzí rice farmers from the conservationists managing Palo Verde National Park. The same ducks that form spectacular flocks, prized by hunters and environmentalists, are destructive pests in rice fields. One study of black-bellied whistling duck gut contents in Palo Verde found that 92 percent of their food was rice.[60] This nocturnal species spends its days resting in protected parklands and stages raids on neighboring fields at night.

I joined Gerardo Mesa, one of Bagatzí's original rice farmers, on a late-night expedition to his rice parcels to learn about the challenges of living in the shadows of a conservation spectacle. Like all farmers who live on the borderlands of Palo Verde National Park, he has to watch his fields every night for fifteen days after planting a new rice crop. The new sprouts are the favorite food of black-bellied whistling ducks. When I arrived at his house late one night, Gerardo was bustling in his back room—he had just taken a quick shower and was collecting his things. He wore a floppy round hat, a long-sleeved shirt, and thick brown pants. Gray stubble was sprouting from his sunbaked brown skin.

While Gerardo's primary goal on this nighttime trip was to chase off ducks, he also told me about complex institutional ecologies and other oblique powers that were slowly moving farmers off their land. Gerardo talked about the challenges of living with a diversity of parasites. Many landless Latin American farmers in the 1980s, according to Peter Brown, were "stuck in the middle, between demands from parasites like intestinal flatworms for a portion of their dinner and demands from macroparasites, the landowners" who demanded rent.[61] Since the time of Peter Brown's work, patterns of landownership have shifted in some parts of Latin America. Farmers living on the margins of Palo Verde National Park claimed title to their own fertile plots of land in the 1980s, but tenacious micro- and macroparasites have not gone away.

Traveling along the borderlands of Palo Verde with Gerardo, I learned about multiple species of parasites that were interrupting the economic system of rice production and destroying the livelihoods of farmers. Emergent diseases, crop pests, and unruly forms of life were jumping around, exploiting opportunities in novel circumstances. "Some of the worst weeds," said Gerardo, "are a lot like rice [arroz] itself—they are called arrozon [rice weed] and arroz pato [duck rice]. These plants grow taller than the rice and shade it out." Later I learned that arroz pato is the exact same species as cultivated rice (Oryza sativa), while arrozon is a closely related member of the same genus (Oryza latifolia). Clever tricksters, these plants have found a vulnerability within the machinery of industrial food production. Changing form, they have become other. They are rice but not quite right.[62] Both of these weedy forms grow faster than the cultivar varieties, producing seeds before harvest time. They accumulate in the seed banks of the paddies, reproducing with every crop cycle, gradually taking over whole parcels. Tractors and harvesters can also serve as vectors for transporting these parasitic plants from field to field.

Bouncing along the dirt roads leading to the rice parcels of Bagatzí in a black Isuzu pickup, a community-owned vehicle that was clinging to life,

FIGURE 9.10. Black-bellied whistling ducks rest in Palo Verde National Park during the day and stage raids on the nearby rice fields of Bagatzí at night. This photograph by Hans Hillewaert was taken in Caño Negro, Costa Rica.

Gerardo also told me about the habits of the waterfowl that were visiting his fields. "Black-bellied whistling ducks are hard to scare," Gerardo said. "Swarms of them just move through the freshly planted fields, gobbling up the rice seeds." Anyone who plants crops during the onset of the dry season, in January and February, must stay awake all night to contend with clouds of ducks. "If you fall asleep for fifteen minutes your crop is lost," Gerardo said. We stopped in a parcel belonging to his brother that had just been planted with rice days earlier. Makeshift lanterns, glass Welch's grape juice bottles filled with kerosene and a wick, were spaced along the terrace walls surrounding the paddy. His brother had been out there earlier in the evening to light them. "It's beautiful out here," Gerardo said, pointing up at the waning moon emerging on the horizon. "The ducks see the lights that we've put out here and think that there is a village."

As we chatted about environmental injustice—about the money that was being spent by U.S. duck hunters to create habitat for crop pests— Gerardo pointed me toward other lines of flight. He gestured to global po-

FIGURE 9.11. Gerardo Mesa surveying his brother's rice field.
Photograph by Eben Kirksey.

litical and economic forces that were exposing human bodies and wetland
ecosystems to a host of dangerous chemicals. If Latin American farmers
had once been "stuck in the middle, between demands from parasites
like intestinal flatworms for a portion of their dinner and demands from
macroparasites, the landowners," the situation was now more complex.[63]
Gerardo, like most Bagatzí residents, was a landowner. But he was being
squeezed by new species of macroparasites—national insurance schemes
as well as multinational corporations like BASF, Dow, and Monsanto.
With multiple species of microparasites feasting in his fields—bacteria,
fungi, worms, mites, and flies, as well as arrozon—he was buying a costly
diversity of fungicides, herbicides, and insecticides.[64] The residents of
Bagatzí were falling into debt. Farmers like Gerardo had become hope-
lessly entangled, seemingly with no way out.

Bagatzí rice farmers use Talcord, produced by the German company
BASF, to kill little *gusanos*—worms that eat rice seeds as they are ger-
minating. Talcord is also used to protect the rice crop at a later stage
in the cycle—when *chinches*, insects in the order Homoptera, arrive in
droves, sucking at the stems or leaves of the rice plants. I happened
to find an empty bottle of Talcord discarded on the side of the road.

"DAÑINO!" (Harmful!), the label warned. "Antidoto: No Tiene" (Antidote: Not Known). The label also noted that Talcord is toxic for fish, bees, and crustaceans.

A diversity of other *plaguicidas* (fungicides, herbicides, and insecticides) are used to combat other organisms: the mites, flies, bacteria, and fungi that are *plagas* (plagues or pests), attacking rice crops. One study in the neighboring village of Tamarindo found nine herbicides, five fungicides, and nine insecticides in use.[65] Many rice paddies drain into the national park, and as a result agrochemicals are entering protected lands. The impacts of these fertilizers and poisons are difficult to track and understand. One study of aquatic macroinvertebrates—insect larvae, snails, and isopods, among other creatures—found more abundance and diversity of organisms in the runoff from the rice fields than in the open waters in nearby protected wetlands. The relative lifelessness of the wetlands was attributed to low oxygen levels. A conference paper, based on a literature review, suggested that agrochemicals might be altering the endocrine system of fish in Palo Verde.[66] In my own literature review, I was unable to find any studies of the impacts of these chemicals on the humans who live in Bagatzí. A government doctor, who visits Bagatzí every fifteen days, told me about cases of serious exposure to agrochemicals—resulting in central nervous system effects and states of unconsciousness.

Pointing out a field of sugar cane next to where we were standing, Gerardo told me about how this plant was slowly pushing rice farmers off their land. "Cane fields are refuges for rice pests, like rats. During the day whistling ducks hide in the sugar plants and attack the rice sprouts at night," he said. Sugar cane companies, the Taboga Refinery and the Old Sugar Factory, were approaching rice farmers and presenting quick fixes to their financial woes. Offering six-year contracts to landowners, the cane companies were taking on their debts and making payments to the banks where loans were owed. The companies were taking care of everything. Sending in crews of cane workers, usually migrant laborers from Nicaragua, the companies were doing all the planting, the chemical spraying, and the harvesting. With no work left to do, with their fields tied up in the financial schemes of outsiders, the rice farmers of Bagatzí were also becoming nomadic—departing for opportunities further afield.

Certainly sugar—like rice plants and jaragua grass—has become a keystone species driving the emergence of novel ecosystems. Fernando Ortiz regarded sugar as an agent of transculturation in Cuba, as a force driving cultural loss, acquisition, and innovation. The character of sugar is "white, sweet, and odorless," in the words of Ortiz, "nutritive, arousing, joyful, pleasing to the flesh, sensual."[67] Some Bagatzí residents were learning to live with sugar, planting this agent of transculturation and ecological emergence on their own terms. Other residents, like Gerardo Mesa, were resisting sugarcane, a plant that was spreading through the hinterlands of Costa Rica with literal and figural rhizomes. In their resistance, many residents of Bagatzí were caring for other organisms whose bodies had been possessed, and then abandoned, by the spirit of capitalism.

Gerardo Mesa told me how concrete hopes and livable futures have been tied to the continued presence of cattle and capitalism in Palo Verde National Park. Purchasing cattle, the ur-companions of capital that were once evicted from the park, has surprisingly enabled some Bagatzí farmers to wander through the protected reserve. While cattle were not effective in the killing of cattails, they proved integral in managing jaragua, the flammable grass with animal rhizomes, which had grown out of control in tall impenetrable stands. Jaragua prevented the forest from regenerating in the early years of the national park by shading out tree seedlings. Short on funds and failing to find workable mechanical methods for controlling invasive grasses, authorities were letting small herds of cattle run through the parklands like living rivets. By early 2009, the time of our late-night foray into rice fields, this reforestation program had become a success story: "In all of the park, there are now only small patches of jaragua," according to Ulises Chavarria, a plant taxonomist who became director of the national park. Cattle had become a surrogate species for the regenerating dry tropical forest in Palo Verde, an ecosystem that earlier generations of bovines had destroyed.

Light grazing by cattle in Palo Verde National Park reduced the risk of fire and allowed tree seedlings to emerge from the shade of the tall grasses. As cattle munched away at unwanted rhizomes, they entered into related stories of interspecies surrogacy. Dan Janzen, a renowned ecologist from the United States, understands these cattle as surrogates that might replace huge mammals—like extinct relatives of horses, ground sloths, and gomphotheres (mastodon-like creatures)—that disappeared in the late Pleistocene (about 12,000 years before present). Many tropical tree species with large fruits were orphaned by these extinctions—they had depended on the Pleistocene fauna to distribute their fruits. Cattle and horses, Janzen demonstrated, are capable seed dispersers for these specific types of tropical trees. With the introduction of these animals by European settlers, the distribution of plants in the emergent ecosystems of the tropical Americas came to more closely resemble the Pleistocene era than the forest that existed just prior to the Columbian Exchange.[68]

FIGURE 9.12. (*left*) Elizabeth Castro, a resident of Bagatzí, milking a cow that had been grazing in Palo Verde National Park. Photograph by Daniela Marini.

FIGURE 9.13. (*below*) The Mexican calabash (*Crescentia alta*) has a cannonball-sized fruit with a hard shell. Horses, which freely graze in Palo Verde National Park alongside cattle, break open these fruits and help disperse the seeds. Photograph by Eben Kirksey.

FIGURE 9.14. The jabiru stork is one of the most endangered birds in Costa Rica. According to published surveys, the population reached a bottleneck of eight jabiru breeding pairs (Alvarado-Quesada, "Conservacion de las Aves Acuaticas de Costa Rica," 52). While I never caught a glimpse of this endangered species in Palo Verde's parklands, in a single day the Argentinian biologist Daniela Marini counted a total of fourteen jabiru feeding in the rice paddies near Bagatzí. This picture shows a jabiru feeding on fringe-toed foam frogs, and other amphibians, as a fangueo tractor plows the rice field of Gerardo Mesa's brother. Photograph by Delphine Farmer.

Hopes are sometimes grounded in the arrival of events or figures on future horizons. At other historical moments, or in different cultural locations, hopes are pinned on the anticipated departure of dreaded or despised figures.[69] If Ulises Chavarria, the park director, grounded concrete hopes for the future of the tropical forest in the absence of jaragua grass, Gerardo Mesa was pinning his own cautious dreams on the departure of ducks. While initiatives to control jaragua with cattle were successful, various projects to make more ducks by killing cattails proved unsuccessful. The ecosystem that conservationists were trying to fasten in place proved to be unstable. As mechanical rivets broke down—as the fangueo tractors and the papermaking machines went out of commission—an ecological community structured by cattails reemerged.

During our nighttime expedition, Gerardo Mesa and I found evidence— the presence of a relative absence—that fed his modest imaginings about a livable future. We heard a few isolated pairs of ducks on the wing, but none were actively foraging in his parcel. Talking late into the night, we discussed how oblique powers were shifting, how ducks were being pushed out of the parklands, moving away from rice fields. Funds from U.S. hunters, who had been sustaining the "duck factory" in Palo Verde, were drying up amid the global financial crisis. Ulises Chavarria was using some of his own limited funds from the Costa Rican government to crush cattails in the immediate vicinity of Palo Verde Biological Station—creating a vista for birdwatchers and a space for field ecologists to conduct research in the open marsh. But large-scale habitat modification catering to migratory ducks was no longer in his plans.

Returning to Gerardo Mesa's parcel late one afternoon, I found fringe-toed foam frogs calling to each other in rice fields and in the drainage ditches of sugarcane fields, aqueous microenvironments awash in chemical toxins. Following my first contacts with this tenacious amphibian, after working with biology graduate students who failed to make it speak to semiotic theories, I searched for its noisy din in unexpected places. In this landscape that had been blasted by multiple waves of capitalism, I came to understand these frogs as living figures of hope.[70] Fringe-toed foam frogs are listed as being of "Least Concern" on the IUCN Red List of Endangered Species. They are flourishing in a time of extinction, figuring into emergent ecologies. Constantly switching from one element to another, these ontological amphibians are hopping among environmental worlds—running wild in la frontera, the borderlands, where capital is warring with itself in a schizophrenic dance. Living as a parasite in worlds designed for others, foam frogs are generating a noisy cacophony—refusing to speak to anthropocentric concerns, invading and occupying agricultural ecosystems flooded with toxins. The wily nature of the fringe-toed foam frog also offers a potent reminder of the limits of human knowledge and forms of representation. In the words of Jean Rostand: "Theories pass. The frog remains."

CHAPTER TEN

POSSIBLE FUTURES

. . .

Leaving the fringe-toed foam frogs behind in Costa Rica's lowlands, I traveled to the highland community of Monteverde, where I continued studying emergent ecological assemblages involving a multiplicity of agents—capital, cattle, rhizomorphic grasses, earthworms, and convivial tree species. English-speaking farmers and scientists helped transform Monte Verde, a two-word place name (Green Mountain) into Monteverde, a singular destination for tourists.[1] The region was settled in 1951 by peace activists, a group of Quakers seeking refuge from U.S. militarism, who became cheese makers and dairy farmers. Expatriate biologists followed in the 1970s—studying the ecology of cloud forests in lands surrounding the Quakers' pastures.[2] By the 1990s, Monteverde had become an international tourism mecca. "On a good day," according to the Lonely Planet guidebook, "Monteverde is a place where you can be inspired about the possibility of a world in which organic farming and alternative energy sources are the norm. On a bad day, Monteverde can feel like Disneyland in Birkenstocks and a zip-line harness."[3]

Many critical accounts of tourism and conservation in this popular destination have been published. Martha Honey, a veteran reporter who helped expose covert CIA operations and the Iran-Contra affair in the 1980s, found "a mixture of hype and experimentation, superficiality and creativity" in Monteverde when she visited in the 1990s. "Industry promises before international forums and 'green' imaging in slick brochures were juxtaposed, out in the field, with grassroots struggles around national parks and nature tourism by some of the world's poorest and most marginalized peoples."[4] Monteverde is "a watershed, a space of recreation, a space of adventure, an exhibit hall for environmental education, a laboratory, a refuge of spiritual inspiration," and, above all, "a tourism spectacle," in the words of Luis Vivanco, a cultural anthropologist.[5]

FIGURE 10.1. The resplendent quetzal (*Pharomachrus mocinno*), an iconic emblem of conservation initiatives in Costa Rica. Photograph by Dominic Sherony.

One animal in particular, the resplendent quetzal—"a migratory bird with colorful plumage and mythical connotations"—is key to the spectacular relations of Monteverde. Like other spectacles described by Guy Debord, these birds are often viewed with "false consciousness"—illusions about the political and economic relationships that sustain their existence and their ecological communities. Quetzals "mediate social relationships between consumers and producers in a tourism economy," according to Vivanco. Images of quetzals are produced by Costa Rican farmers, hospitality industry staff, tour guides, and local conservationists, who are in a subordinate relationship to consumers: visiting tourists, scientists, and documentary crews. At their most spectacular, quetzals are seen as existing outside "the highly specific socioeconomic conditions that created the possibility of their viewing."[6]

The quetzal was a flagship species, a charismatic animal whose image was used in a marketing campaign in the 1980s to raise hundreds of thousands of dollars for outsiders to purchase land in the Monteverde region. Dozens of subsistence farming families, with homesteads on desired plots of land, were induced to move. Some relocated to other farmlands, while others embraced opportunities emerging with the new economy of tourism. As resplendent quetzals and English-speaking tourist guides flourished, others struggled to survive in the shadows of the spectacle.[7] Hunting was outlawed in the protected areas of Monteverde, as was the gathering of forest plants—like hearts of palm, wild avocados, and timber trees. Wild cats—pumas, jaguars, and ocelots—preyed on the farmers' livestock as they proliferated in this emergent ecosystem.

One woman, whose family willingly sold their homestead to conservationists in the 1980s, moved to town and found employment as a maid in one of the many hotels that sprang up in the region. She befriended me in 2008, after we met via her husband, a day laborer who sometimes cleared trails in the Monteverde Preserve. Steady wages, proximity to family and friends, and an affable boss meant that she was generally upbeat about her position in the new economy. She was the stable breadwinner of the family. But when we passed each other on the street one day, she furtively confided in me: "My nose started bleeding at work today. I think that it is the cleaning products they make me use." Emergent economic opportunities have thus exposed Costa Rican families to new vulnerabilities, new precarious modes of existence.

Biologists from the United States who have become Monteverde residents told me (off the record) that life on earth was in an increasingly precarious position. Pointing to anthropogenic changes in environments at local and global scales, resident ecologists talked about troubling signs. Despite the best efforts of conservationists, endangered species were disappearing from protected reserves, seemingly like rivets being popped out of an airplane wing. In the 1980s the resplendent quetzal often appeared alongside another charismatic animal, the golden toad of Costa Rica, on posters for the Save the Rainforest campaign.[8] This charismatic frog was confined to a tiny environmental world: a strip of high-elevation elfin forest that measured eight kilometers long and only half a kilometer wide at the center of the Monteverde Cloud Forest Reserve. Living within an underground tunnel system between the roots of cloud forest trees for most of the year, it only emerged for five to ten days early in the wet season to breed. Shortly after the ecotourists began arriving by the thousands, the golden toad went extinct.[9]

"The disappearance of the golden toad has coincided with the phenomenal growth of tourism," writes Martha Honey. "The history of the golden toad and that of ecotourism are intertwined, and some speculate that an ecotourist (or perhaps a scientist) may have carried into Monteverde's rain forest an alien organism that caused a plague among the reserve's toad population. If true, it is ironic, since Monteverde scientists and residents have consciously used conservation grants and ecotourism profits to protect the habitat of the golden toad and other exotic, endangered

FIGURE 10.2. Milk cartons and junk mail, featuring missing amphibians in the place of missing children, were created by ecoartist Ruth Wallen in 1996 as part of a digital installation called *If Frogs Sicken and Die, What Will Happen to the Princes?* The carton featured the golden toad of Monteverde, Costa Rica, an animal now presumed extinct. Image courtesy of Ruth Wallen.

species."[10] Martha Honey's essay, "In Search of the Golden Toad," was published in 1999, just months before Joyce Longcore described and named *Batrachochytrium dendrobatidis*, the frog-killing chytrid fungus. While ecotourists may have unwittingly brought chytrid zoospores to Monteverde, and conservationists may have ironically hastened the doom of one of the very animals they were trying to protect, this particular extinction story also contains elements of a broader tragic plot about uncaring human masses. It must be told against the backdrop of global climate change.[11]

An El Niño/Southern Oscillation event hit Central America in 1986–87, resulting in abnormally low rainfall in Monteverde. Marty Crump, a herpetologist who was studying the tadpoles of the golden toad, wrote in May 1987, "This has not been a good year for the toads with respect to recruiting new individuals into the population. During April nine of the ten pools where I found eggs had dried up completely before the eggs had even hatched, and all the hatchlings from the tenth pool dried up later. From May's breeding bout, half the pools containing eggs dried up before the tadpoles were a week old, and only twenty-nine tadpoles survive in the remaining pools."[12] A local Costa Rican naturalist, Eladio Cruz, saw some golden toads shortly after Crump made these observations. This species has never been seen since and is now presumed to be extinct. This particular extinction event in the recent past fueled feelings of anxiety and dread about the near future. Over drinks one night at Moon Shiva, a tourist bar that has since closed, one young biologist with a PhD from Princeton told me, "Articles about the golden toad have been published in all of the top journals—like *Nature* and *Science*. As people advance their academic careers, and raise funds to create nature reserves, there are forces at work beyond our control. To me it is clear the world is ending, so we might as well go out with a bang—drinking and partying!"

HOPE ON THE MARGINS

As credentialed biologists from prestigious universities intimated their apocalyptic dreams and hedonistic desires, I found Costa Rican thinkers and tinkerers who were doing practical and imaginative labor to keep concrete hopes alive. Milton Brenes, the coordinator of a reforestation program at the bilingual Monteverde Cloud Forest School, is working against general feelings of anxiety about the environment to create a shared future with a multispecies community he loves. Milton is a short, mustached man with a seemingly endless reserve of energy and enthusiasm. He is a farmer, an organic intellectual, and a bricoleur. On eleven hectares of derelict pasture in the highlands of Costa Rica, Milton is re-creating a forest in collaboration with a multitude of plants, animals, and students. In parallel, the Cloud Forest School, a private K–11 institution, is endeavoring to generate a community of English-speaking environmentalists to care for the rejuvenating landscape.[13] Skilled at working with emergent opportunities and shifting contingencies, Milton guides a labor force of children and volunteers—ranging in age from preschool to their early thirties. Paying careful attention to their interests has enabled Milton to bring diverse actors together in a common project. Casting my lot with Milton, I joined this initiative as a volunteer who helped plant trees and as a participant observer of reforestation in action.

"Hope for the future," Milton told me in Spanish, "lies in the conservation of diversity." At first blush, this sound bite might sound like a simple repetition of environmental platitudes. Careful attention to Milton's rhetoric and practice reveals instead a clever way of speaking in terms a transnational audience can hear.[14] Hope here is emerging within a contact zone shaped by transnational regimes for regulating life, by elite North American visions of Nature, and by predictions of multiple catastrophes on future horizons. Diversity here is a product of cultural hybridity, *mestizaje*, where a diversity of uses in indigenous traditions, a diversity of articulations with market economies, and a diversity of taxonomic forms all matter.[15] Using the language of hope and diversity helps Milton direct the impulses of short-term foreign staff and volunteers like myself, drawing us into the living architecture of regenerating cloud forest trees. Amid major ecological and economic transformations, he has aligned the interests of heterogeneous life forms, enfolding us together in an assemblage of multiplicities. An ecosystem is emerging, a riotous collection of strangers, which has begun to exceed Milton's own vision. Entrepreneurial plants, worms, and other animals are generating their own multispecies communities.[16]

Rather than focusing his efforts on preserving rare forms of life, Milton has cultivated alliances with robust trees that are helping him generate convivial assemblages. "We are producing twelve different tree species in our reforestation program," Milton says. "This will ensure that our students and tourists can have the opportunity to see important parts of the local fauna." *Inga punctata*, which is a hearty and fast-growing plant often found along roadsides and in abandoned farmlands, is one of Milton's favorite plants. "*Ingas* produce food for monkeys," Milton continues. "We want to attract monkeys to our school since they are dynamic, they are fun, since they are a species that the tourists like." The leguminous fruits of these trees, which are tasty food for capuchin monkeys, also attract parakeets, squirrels, large rodents called agoutis, and relatives of raccoons called coatis. Known as ice cream beans in English, or *guaba* in Spanish, the sweet-white pulp in *Inga* seed pods is a prized recess snack for children at the Monteverde Cloud Forest School. *Inga* flowers are also of interest to a diversity of beetles, butterflies, wasps, flies, hummingbirds, and bats.[17]

Fast-growing *Inga* saplings are quick to send out a spreading crown and shade out cosmopolitan grasses in the abandoned pasturelands. These leguminous plants fix nitrogen in the soil, creating conditions for the flourishing of other trees. *Ingas* can also summon animal familiars to their defense. All *Inga* species in Costa Rica have extrafloral nectaries, which enlist ants like *Ectatomma* as well as wasps and flies into multispecies ensembles (see chapter 2, figure 2.4).[18] In coffee and cacao plantations, *Inga* trees are planted to support an "ant mosaic" that helps protect crop plants from herbivory.[19] Milton plants *Inga* saplings at the Monteverde Cloud Forest School to help build a convivial world, tooth and nail, in concert with others.[20]

FIGURE 10.3. Milton Brenes's center for tinkering and thinking with plants was the greenhouse, a small structure with orchids, ferns, and bromeliads twining around each other and dripping from the ceiling. Here seeds were germinated, seedlings were nurtured, and rescued epiphytes were nursed back to life. Plantlets thrived on the floor, benches, and tables. Photograph by Eben Kirksey.

Conservation textbooks suggest planting tree species with attractive fruits to "encourage seed dispersal to the site via frugivores" (fruit-eating animals).[21] Each visit from a bird, an agouti, a monkey, or a bat to fruit trees brings the possibility of new plants sprouting from seeds in poop left behind. Pushing past textbook approaches to reforestation, Milton works to recruit allies—humans from disparate social worlds and multiple other species—to build an expanding network. In other words, he is always trying to multiply his forces with a multitude of entrepreneurial agents, seeking to generate an ever-expanding project of interessement, of enlistment.

"We started the reforestation project in the year 2000 and in total we have planted 13,630 trees," Milton told me. Kindergarteners and preschoolers put in a few important tree species in the first year. When Milton introduced me to the project in 2008, just eight years after starting, scores of trees were already towering over his head. By leveraging the interests of animals from a range of locations, by drawing allies into the generative processes of pollination, seed dispersal, and propagation, the *Ingas* were already helping create a multispecies community. Assimilating others into themselves, destroying this otherness while drawing sustenance, these plants began weaving insects, rodents, primates, and birds into emergent ecologies on the school grounds.[22] Another convivial tree, hollow heart (*Acnistus arborenscens*, Solanaceae), was already attracting a wide variety of insect pollinators. Wasps, flies, butterflies, and beetles were being drawn in by a rare fragrant compound called orcinol dimethyl ether, which was almost undetectable to my own human nose. The corky bark of the hollow heart trees also proved to be an ideal substrate for growing epiphytes, plants that grow on the branches of other plants (*epi*-, upon; -*phyte*, plant). After just eight years, one hollow heart tree at the Cloud Forest School was full of orchids (*Pleurothallis* sp.), bromeliads, mosses, lichens, piper plants (*Peperomia* sp.), and melastomes.[23] Students were snacking on the tasty fruits, which are also attractive food for bats and over forty species of birds.[24]

Milton's reforestation program, and other educational initiatives at the Monteverde Cloud Forest School, have been chronically underfunded. The Nature Conservancy initially loaned the school $189,862 to purchase the pasture land in 1992 from a wealthy Costa Rican family.[25] After charging the cash-strapped school 8 percent annual interest on this loan for several years, the Nature Conservancy demanded a payment of $270,000 by a June 2000 deadline.[26] Administrators met this deadline, at the eleventh hour, but have been struggling ever since. A fund-raising campaign in 2008 sought to raise the wage of the Costa Rican and U.S. expatriate teachers—who were being paid less than $3.50 an hour—and to provide more scholarships to students, who were over 90 percent Costa Rican. As a result of the school's fragile finances, Milton was running the reforestation program on a shoestring. Despite day-to-day difficulties, the lack of foreign funding was liberating in a sense. Disinterest from the Cloud Forest School Foundation—a 501(c)(3) based in Sewanee, Tennessee—meant that Milton had the freedom to implement his own vision of reforestation while being attentive to local needs and interests. Milton was not working outside the machinations of global capital, but was cultivating diversity at the margins.

Milton's understanding of diversity had been shaped by his earlier work on projects with the Monteverde Conservation League.[27] Rather than focusing on biodiversity protection concerns of biologists, the Monteverde Conservation League was oriented to "the needs, interests, and perspectives of rural communities and small-scale dairy and coffee farmers."[28] Initially, exotic species such as casuarina (*Casuarina equisetifolia*) and cypress (*Cupressus lusitanica*) were planted because they were fast growing, known to foresters, and on a Costa Rican government list of trees approved for financial incentives. As thousands of ecotourists began arriving in the 1980s, intent on seeing spectacular quetzals, toucans, and monkeys, economic opportunities emerged in ecosystems forested with local trees. The Monteverde Conservation League shifted their focus from exotic trees to native plants amid pressure from farmers who understood local plants as "more ecological, more productive, and more useful," according to Milton. By the early 1990s, Milton and the other League personnel were growing forty-nine species—six kinds of exotic trees and forty-three natives.

Working with resources that were ready at hand, the Monteverde Conservation League had developed simple reforestation methods. Milton used similar methods at the Cloud Forest School. Periodically he led the youngest students into the forest where they collected seeds from under *patrónes*, large "patron" trees that provided stock for the project. The seeds were brought back to a greenhouse where they were soaked in water for twenty-four hours—to kill insect larvae, eggs, and worms—before being laid to rest on three different beds for germination. One bed had leaf litter from the forest, one had decaying wood, and a third had soil. "Seeds of some plant species won't germinate if they are covered by soil," Milton told me. "Many trees in the primary forest need the specific moisture levels found in leaf litter or decaying wood to germinate." Once plants were big enough to leave the germination beds, they were housed in reused milk and juice cartons—collected, washed, cut to the proper specifications, and packed full of soil. Using found objects and organisms—gleanings from the detritus of industrial food production and the litter of leaves in the forest—Milton was re-creating an ecosystem.

THE RHIZOSPHERE

Despite pressure from some influential local biologists to use only native organisms, Milton brought a multitude of undocumented aliens, laborers rumored to be from California, to help with his reforestation project.[29] These workers were earthworms with a talent for converting food waste, the lunch leftovers of the teachers and schoolchildren, into a nutritious microbial assemblage that fed plant roots. Paul Sataltan, a Peruvian biologist, first brought the ancestors of these worms to the Monteverde region in 1997 from a technological institute in the lowlands of Costa Rica. Originally the worms were imported for different work: to deal with the waste, the fruit pulp, left over from the processing of coffee beans at the Café Monteverde Cooperative. Given as gifts to neighbors, conserva-

FIGURES 10.4 AND 10.5. Planting trees at the Monteverde Cloud Forest School in 2008. When I returned to the same site in 2014, a young forest, with an intact canopy, had reemerged in the barren ground. Photographs by Eben Kirksey.

For us to arbitrarily decide who should live and who should die in this complex system is a bit presumptuous. Although some controls are suggested, this chapter's major purpose is to provide a better idea of what you can expect to find. Don't be alarmed at what you see. Information tempers fear. You may even decide to learn more about those critters that you once used to squash when you found them.

FIGURES 10.6 AND 10.7. The convivial community of the worm bin comes to order with little help from humans. Cartoon courtesy of Mary Appelhof and Mary Frances Fenton.

tion institutions, and farmers, these earthworms came to populate local compost bins throughout the region.[30]

Members of an alien species thus began serving as surrogates for native plants in the regenerating cloud forest.[31] Milton started using worms to literally reassemble the forest from the ground up. Each tree seedling was planted in a small hole with a shovel full of material from the worm bin. Heavy grazing by cattle had compacted the lands inherited by the Cloud Forest School, turning the soil into a dense and distorted mass with reduced porosity and permeability.[32] Soil is usually a dynamic, living entity. The forest rhizosphere, the zone of soil around plant roots, is usually teeming with bacteria, microfauna, and symbiotic fungi. These companions are fed with nutrients by the roots and, in turn, they help the plants flourish.[33] Parts of the school's soil had turned into an impoverished desert, with few plant companion species.[34] Karen Masters, who laid plans for the Monteverde school's reforestation program in 1998 shortly after finishing her PhD in biology, told me, "This spot was wind-swept. The ground was like concrete."

Compost helped enliven this soil. Millions of microorganisms and thousands of small burrowing animals were present in the material generated by Milton's composting system. Worm castings usually contain fairly low amounts of nutrients. Yet plants grown in this chemically poor material still often have higher yields than those grown with high concentrations of commercial fertilizers.[35] Symbiotic organisms found in compost promote growth by increasing the bioavailability of nutrients to plants. Soil organisms can also play a defining role in the emergence of novel ecosystems by hindering, rather than helping, plant growth. Mature soils contain pathogens that limit growth of dominant trees while indirectly helping rare plants. Homogenous stands of plants, clusters of individuals that are all the same kind, often attract swarms of nematode worms, insect larvae, and deadly microbes that feed on roots. Attacks on abundant plant species thus create opportunities for others to emerge.[36]

Convivial communities in the rhizosphere, unruly assemblages that foster life for some and hasten the death of others, can do their work in the world with little help from humans.[37] Worm compost bins require little care. Shade them from the sun, keep the compost moist, periodically add more food scraps, and worms will work their magic. "C" for Compost, in the 2005 edition of the *Encyclopedia of Soils in the Environment*, describes the mystery surrounding these multispecies interactions in dry prose: "Compost can have a range of positive effects on soils, microbial communities, and plants. The complexity of these interactions and the diversity of compost types and behaviors pose continuing challenges for both scientific understanding and practical management."[38] Creatures lurking in the cosmos beyond the narrow purview of our understanding are thus exploring possible political articulations among themselves, exploring modes of coexistence in common worlds that are largely imperceptible to humans.[39]

Worms Eat My Garbage, a book I found in a local Monteverde library, contains a cartoon by Mary Frances Fenton that illustrates a Worm Bin Recycler's Association that has crawled into cosmopolitical association of its own accord. Here beetle mites, mold mites, millipedes, sowbugs, springtails, roly-poly pill bugs, mold, bacteria, and pot worms have come together in an assembly that is being called together by the red worm (*Eisenia fetida*). According to *Worms Eat My Garbage*, "Your worm bin is not a monoculture, but a diverse community of micro and macro-organisms that are interdependent. No one species can possibly overtake all the other species present. They serve as food for each other, they clean up each other's debris, they convert materials to forms that others can utilize, and they control each other's populations. For us to arbitrarily decide who should live and who should die in this complex system is a bit presumptuous."[40]

THANATOS AND EROS

While the compost bins of worms were treated with laissez-faire policies, where benign neglect and disinterest allowed a multitude of beings to flourish in the soil, the work of reforestation above ground demanded more careful attention to matters of life and death. Since the Monteverde Cloud Forest School became a conservation easement, strict rules from Costa Rica's environmental ministry theoretically governed the management of killing and fostering life on the school grounds. But, in actual practice, local environmental ministry staff were up to their ears in other issues and rarely had time to visit. Milton was given a mandate by the state to determine which organisms should be cultivated and which should be killed on the school grounds. Milton became the benevolent *oikonomos* (the head of the estate) with the power to determine which plants deserved protection and which might be stripped of the right to life.[41]

Heroes and villains in this project of reforestation did not exist in any absolute sense, but were judged on the basis of situated, contingent action and effect.[42] As Milton began to care for an emergent forest, he began making high-stakes, and potentially arbitrary, distinctions between species he considered the enemy and species he regarded as allies.[43] Deborah Bird Rose argues that "death and continuity are core aspects of the integrity of life." Caring for others involves watching out for their interests, defending against outsiders or interlopers. Rose calls for us to embrace the necessary labor of killing in a way that affirms and sustains multispecies connections.[44] Milton applied similar norms about death and life to the plant communities under his watchful eye. Singling out enemies that were disrupting diverse connections, Milton worked to kill some plants in a way that twisted death back into life.[45]

East African star grass (*Cynodon nlemfuensis*) became the number one enemy of the Cloud Forest School's reforestation program. Like jaragua grass, which was introduced to lowland pastures of Costa Rica, this plant has long been a companion of cattle in Central America's highlands. Introduced to the region in the 1970s as a plague of spittlebugs and froghoppers attacked the fields of local dairy farmers, star grass was once loved by many people in Monteverde. Star grass was less nutritious and less palatable for livestock than other grasses but became the preferred plant because of its resistance to insects. The same fortitude that made dairy farmers love this grass led Milton to hate it. Unlike the California red worms, allied aliens that were helping him build new worlds, he judged that star grass had become irredeemably destructive. Abandoned fields inherited by the Cloud Forest School were overgrown by tall and dense stands of star grass that shaded out other plants. Attempts to chop down these thickets with machetes or motorized weed whackers were frustrated as this tenacious plant quickly resprouted.

Star grass has deep roots, rather than rhizomes. It can spread vertically, as well as horizontally, with long stolons, runners that weave among the branches of emergent trees, weighing them down and blocking the sunlight.[46] While Milton's trees struggled to form links with a multitude of beings in the world, the star grass was more autonomous. This plant had a limited sense of self. An unrelenting commitment to self-preservation and self-propagation once made star grass an ideal companion of cattle as well as capitalist patterns of production and consumption.[47] These same properties made it the enemy of reforestation. Levinas has written of beings for whom "interest takes dramatic form in egoisms struggling against one another." For such beings, "*esse* is *interesse*; essence is interest."[48] Combating this vegetal egoism, Milton formed alliances with entrepreneurial plants with more convivial approaches to interessement.

While star grass had egoistic struggles with spittlebugs, froghoppers, and other insect herbivores, fast-growing *Inga* saplings began shading them out with their spreading crowns while summoning a diversity of insects to their aid. Surprises emerging within the abandoned pasture also prompted Milton to develop new tactics and strategies. An aster bush, known only as *monte* (a word meaning "mountain, backwoods, wilds, bushy scrub, or weed"), grew of its own accord within the tangled mats of star grass. Monte bushes began carpeting the land with light blue flowers and thousands of tiny seeds.

Milton began leaving the patches of land with *Monte* alone, watching as this native bush began to gradually displace the African grass. Vegetative dynamics among monte bushes and star grass were eventually overshadowed by trees he planted nearby as well as saplings spontaneously growing within. Costa Rican fiddlewood trees, *Citharexylum costaricensis*, began emerging within the fray. Likely arriving in the pasture from seeds distributed by birds pooping on the wing, the fiddlewood seedlings began secreting nectar from glands on their underside, summoning ants and

FIGURE 10.8. A bush called monte spontaneously emerged within the abandoned pastures and helped Milton rein in the wild roots of star grass. Photograph by Eben Kirksey.

other helpful insects to their defense. Quickly growing and producing flowers, they attracted small moths at night and stingless bees during the day. With fruits eaten by more than twenty species of birds, including toucans and turkeys, the fiddlewood trees helped open up opportunities to a multitude of others.[49] Responding to historical contingency, Milton was welcoming some emergences and squelching others. Always on the lookout for new talented or useful species, he was improvising at the intersection of multiple social worlds and ecological communities.

COMPETING VISIONS OF NATURE

When I returned to the Monteverde Cloud Forest School in February 2014, Milton brought me to places where I had helped plant trees and pull out weedy grasses. Exposed sections of the school grounds, once choked with star grass and underbrush, had become covered by a shady forest canopy. Milton pointed out the trees that were starting to overshadow the classrooms. There were *dama* (*Citharexylum costaricense*) as well as *Lorito* (*Cojoba costaricensis*), both which were producing fruit for migratory birds and squirrels. Nearby students were snacking on the fruits of a

FIGURE 10.9. A tree locally known as Maria (*Conostegia xalapensis*) is commonly found along roadsides and in abandoned pastures in Monteverde. The small dark fruits are favored by a diversity of birds as well as schoolchildren. These two students of the Cloud Forest School were enjoying a snack of Maria fruits from a branch given to them by Milton. Photograph by Eben Kirksey.

large *Conostegia xalapensis* tree, a member of the melastome plant family. Other children showed me their favorite tree, *mocos de mono*, or monkey boogers (*Saurauia montana*). The small lobed fruits of this tree contain a tasty sweet slime. "But don't eat too many monkey boogers," one boy warned me in English. "It makes your tongue and throat all tingly."

We were sitting in a park that Milton had created, on a bench he had built with help from students. "This park is a place where students and volunteers like you can come back to see the results of your efforts," he said. Signs labeling the principal tree species in his reforestation program gave the park an orderly feel. Milton said that he had been meticulously caring for this space, establishing well-defined trails, keeping the area free from weeds and human trash. This was clearly a space for people, for learning about trees and their uses. "Over here we have the *chancho blanco* (*Beilschmiedia costaricensis*)," Milton said, "which is desirable for hunters because its seeds are food for some mammals, like the paca (*Agouti paca*) and the wild pig. The chancho blanco is also good for lumber." The park was also becoming a site for encounters with Monteverde's charismatic wildlife. Howler monkeys had been visiting the school. As we talked, a toucan flew overhead and the metallic honk of a bellbird sounded in the distance.

While Milton showed me the trees he had planted with diverse local uses, he also talked about preparing students to engage with economic opportunities that had emerged with ecotourism in Monteverde. A few students, like Milton's own son Johnathan, had managed to use their education at the Cloud Forest School to pursue university education abroad. But most graduates were finding employment locally in the tourism sector—as taxi drivers, hotel receptionists, waitresses, bartenders, and guides. With wealthy tourists whizzing through treetops in zip-line harnesses on land immediately adjacent to the school grounds, students were indirectly learning uneasy lessons about social inequality as they studied natural history in the emergent ecosystem outside their classrooms.

Some expatriates living in Monteverde—biologists and conservationists from the United States—were critical of Milton's reforestation efforts, claiming that he was not doing enough to enhance the lifeways of certain charismatic species. These critics pointed me to the research of George Powell, who had been conducting radiotelemetry studies of resplendent quetzal migrations in the region. Powell was part of the team that petitioned the Nature Conservancy to help buy the land in 1992, writing that this site was "very important to the conservation of the quetzals because it is situated adjacent to the Monteverde Preserve, but at a lower elevation which is heavily utilized by migrating quetzals." Forest fragments near the farm had more than fifteen species of wild avocado trees, in the Lauraceae family, "all the important ones for quetzals." Powell concluded that the lands that were eventually acquired by the Cloud Forest School could "provide a bridge for migrating quetzals and a critical source of food in this altitudinal zone."[50] Critics claimed that Milton was not doing enough to help the quetzal and other birds. They charged that he was not planting enough wild avocados.

When he began the reforestation program in the year 2000, Milton planted many Lauraceae seedlings. But many of these fragile plants died. "The only way most wild avocados can advance is with help from other planted trees that are faster in their growth," he told me. Milton has had success in planting only one kind of wild avocado, *Ocotea whitei*, which is known in Spanish as *ira rosa*. In 2008, Deborah Hamilton, a U.S. national who has devoted much of her adult life to the study and conservation of the three-wattled bellbird in Monteverde, told me that *Ocotea whitei* is a "trash tree." According to Deborah, another species in the same genus, *O. monteverdensis*, is an endemic tree that is "preferred by everybody," namely toucans, bellbirds, thrushes, and quetzals. On

FIGURES 10.10 AND 10.11. The Cloud Forest School is located next door to Monteverde Canopy Tour, which was started in 1994 when the tops of forest trees were suddenly opened up to the paying public. A headline attraction is the Tarzan Swing, a harnessed free fall at the end of a long metal cable. Participants are also given the opportunity to climb up entangled fig roots, rope ladders, and guide wires. Photographs by Eben Kirksey.

one occasion, Deborah saw some eight quetzals sitting on a single tree. When *O. monteverdensis* is planted in combination with *O. floribunda*, it becomes a "super quetzal and bellbird tree," she added. These two *Ocotea* species fruit every other year, providing a good seasonal source of fruit that draws the same individual birds back to the same place. Deborah was running a tree nursery where she was nurturing *O. monteverdensis* saplings. In addition to producing trees with fruit for her favorite birds, she told me, her nursery was aiming to match the natural abundance of tree species in forests.[51]

Key questions should be asked of all reforestation projects: "What counts as nature, for whom, at what costs?"[52] The community of expatriate conservationists and biologists in Monteverde is deeply divided with respect to answers. Alan Masters, a biologist who served as the president of the Cloud Forest School Board for eleven years, is critical of reforestation programs that favor charismatic megafauna. "People only think that a plant is good if it feeds something like a bat, a bird, or another animal," Alan told me. "Why make a giant zoo?" he asked. "No one is planting trees for millipedes." While interviewing Alan Masters, I brought up Bruno Latour's proposal for manufacturing new speech prosthetics, for bringing new nonhuman voices into the domain of human representations. I said that our knowledge about plants and animals is constrained by university institutions and the history of particular disciplinary traditions. Alan picked up on some of Latour's arguments right away, rhetorically asking, "How much is the critter like you? How well is it studied? By virtue of design or neglect, who are we doing favors?"

Pulling a book off his shelf, *Costa Rican Natural History* by Daniel Janzen, Alan told me that he had been reading it against the grain. The milk tree, a lowland species that grows to a towering height of fifty meters, has a variety of uses (thus the scientific name, *Brosimum utile*). The sap, a white latex fluid, is drinkable, and the sweet pulpy fruits are edible. The wood is useful in construction, and the thick bark made for a warm and flexible blanket for Amerindian groups. The milk tree is the most abundant canopy tree on well-drained slopes and ridges in Corcovado, a peninsula in southern Costa Rica that is regarded by conservationists as the most pristine region in the country.[53] When Sir Frances Drake visited the region in the sixteenth century, he saw a wilderness, according to Alan, rather than a place with a high density of useful trees. Pointing me to other sites in Central America, Alan said that even after some two thousand years of regeneration following the collapse of the Mayan empire, the forest still shows signs of human presence. There are still higher densities of trees with big juicy fruits and edible seeds near Mayan ruins, when compared with forests farther from these settlements.[54]

Bringing our conversation back to reforestation at the Monteverde Cloud Forest School, Alan asked, "If Milton is planting twelve species that favor particular birds and mammals, what is this doing to ecological dynamics? How will decisions taken today play out over the next two thousand years?" Advocating for a disinterested approach to reforestation, kindred to the laissez-faire policy that allows the Worm Bin Recycler's Association to crawl to order on its own, Alan Masters told me that reforestation projects should be focused on creating spaces for nonhumans. Rather than creating a utopia for bird enthusiasts, or for tourists who want to see monkeys, he thinks we should also focus conservation efforts on mushrooms and bacteria. Alan would like to see more places left alone to see what might emerge. "Reforestation isn't an inherently natural process," he said upon reflection. "With ecosystems there are unpredictable outcomes," he added. "There are emergent properties that we can't anticipate."

After hearing the concerns of Alan Masters and Deborah Hamilton, I went back to Milton and invited him to respond. Walking with me around the school grounds, he recounted his work with emergent properties of plant communities, the experimental interventions he made against the backdrop of unpredictable outcomes. He showed me places where the helpful aster, monte, was still present, entangled in a life-and-death struggle with star grass. He asked me to take a second look at places for humans that had become focal sites of his attention and care (*cuidado*), and other spaces that he was regarding with benign neglect (*descuidado*). In the orderly park, where Milton had established well-defined trails and clearly labeled trees, he was caring for plants with roles in Costa Rican communities. Many of these trees were valuable as lumber or had fruits and nuts that were attractive to pigs, agoutis, and other animals favored by local hunters. Relationships between students and plants useful in mestizo farming traditions were being cultivated here, in an outdoor classroom. Encounters with insects, birds, and bats—all creatures with critical roles in the reemerging forest ecosystem—were also being facilitated in this carefully managed space. In this park, Milton was enacting an ethics of care based on sustained relationships.[55]

Caring for other spaces with benign neglect also enabled Milton to decenter his social world, allowing other forms of life to flourish outside of utilitarian calculations. While cultivating sustained relationships with some plants and allied animals, Milton tactfully kept his distance from others—like the worms in his compost bins and monte, the blue aster. With competing needs and interests intersecting in this regenerating forest—those of small-scale farmers, local hunters, capricious visitors, foreign donors, tourism promoters, and biological scientists—this place exemplified the "contaminated diversity" described by Anna Tsing. "Contaminated diversity is everywhere," writes Tsing. "For better or worse, it is what we have. . . . Diversity continues to emerge, even in ruins." Milton was fostering a diversity of links to economic systems and overlapping ecological communities. Working in the aftermath of environmental destruction, with the detritus of industrial food production, with weedy plants and alien helpers, he was generating lively possibilities.[56]

SURVEYING DIVERSITY

Eladio Cruz, a renowned Costa Rican naturalist, worked with me in February 2014 to conduct a biodiversity survey on the grounds of the Monteverde Cloud Forest School. He also helped me better understand how each plant species we encountered had a diversity of articulations and entanglements in social and biological worlds. Meeting at Eladio's house, at the base of a steep hill leading up to the school, we chatted about how local dairy pastures were gradually being reclaimed by trees, even in the absence of active reforestation programs. A light rainy mist, what Costa Ricans call *pelo de gato* (cat hair), was falling all around us as we climbed the hill. We stopped to look at a tall *Inga* tree and a cluster of mature Maria trees (*Conostegia xalapensis*) growing along the roadside on their own accord. As we entered the school's land, as we began photographing and identifying trees, we bumped into Eladio's son, José Andreas, a high school junior. Walking past the areas frequented by students, we ventured onto a steep hill, into a stand of trees I had helped plant in 2008, just six years before (see figure 10.4). Underfoot was much leaf litter and some plastic litter. Overhead was a canopy formed by lanky *Inga punctata*, guayaba, and ira rosa (*Ocotea whitei*) trees. But Eladio and I were not counting or identifying those species that had been planted as part of the reforestation project. We were instead searching for emergent seedlings and saplings that had germinated on their own.

Star grass was still covering some parts of the hill in an impenetrable tangle. Other parts of the hill had an open understory, with the blue aster, monte, growing sparsely, close to the ground. Eladio identified most plants by sight, occasionally crumpling a leaf to see if it had a distinct odor, or turning it over to see if it had spines or hairs. Rolling the syllables gently off his tongue, Eladio patiently waited for me to scribble in my notebook, recording Latin words and plant families that invoked distant lands and long-dead botanists: *Montanoa guatemalensis* (Asteraceae), *Diospyros hartmanniana* (Ebenaceae), *Sideroxylon puertoricensis* (Sapotaceae), *Beilschmiedia brenesii* (Lauraceae). Repeating variations of familiar stories, Eladio told me about the animals attracted to particular plants: "This one has small seeds distributed by doves and the blue-crowned motmot. That plant bears the favorite fruits of bats. This tree is useful for timber and attracts wild pigs, squirrels, and agoutis."

Walking to the other side of the school, toward the park Milton had created, we passed by a fragment of old-growth forest and by some of the very first trees planted at the school. Two "super bellbird" seedlings, *Ocotea monteverdensis* and *O. floribunda*, were growing spontaneously near a "trash tree" seedling, *O. whitei*, the plant derided by Deborah Hamilton for having substandard fruit. As Eladio identified another plant in the same genus, an *Ocotea tenera* sapling, he brought out a magnifying glass to show me some miniscule domicia, "small houses," on the underside of the leaf. We could see tiny arthropods—perhaps mites or small beetles—scurrying around, but they were too small for us to identify.

Eladio was pointing to surprises beyond the scope of Milton's vision, wild emergences involving multiple species that had come into being amid the play of diverse market economies, pedagogical initiatives, and scientific enterprises. Oblique powers and opposing forces were at work in these multispecies worlds. Allies in Milton's reforestation project—convivial trees attracting animals involved in pollination, seed dispersal, and plant protection—had multiplied their forces to the point where their existence was no longer contingent on humans, worms, or other surrogates. Enfolding each other into an emergent ecosystem, they had brought a multitude into being. Not counting the twelve tree species that had been actively planted, Eladio identified a total of thirty-eight tree species emerging within the reforestation project.[57] These species had not figured directly into Milton's vision for the forest; they were the unexpected figures of hope he had expected, textbook examples of seed dispersal at work.

DISASTER AND OPPORTUNITY

Eladio and I took a break after our biodiversity survey, to eat a picnic lunch on a bench in the manicured park that Milton had created. As we were talking, Eladio sprang up. A tree had caught his eye from across the

other side of the park. Bright-red fruits were hanging down in conspicu-
ous clusters from the branches of a bitterbush (*Picramnia antidesma*), a
shrub that he had never seen up at this altitude. This central part of the
Monteverde Cloud Forest School lands sits at about 1,500 meters, and
Eladio had seen this plant previously only in the nearby San Luis valley
at an elevation of about 1,100 meters.[58] Milton lives in San Luis, and we
later learned that he had brought its seeds up to the school in a modest
attempt to prepare for climate change.

Working under Milton's direction in 2008, I had helped plant lowland
trees on the school grounds that had diverse articulations to market
economies, uses in mestizo traditions, and possible futures in techno-
scientific dreams. Costa Rica's national energy and telecommunications
company, ICE, donated seedlings of valuable timber species—cigar-box
cedar (*Cedrela odorata*), cocobolo (*Dalbergia retusa*, Fabaceae), and false

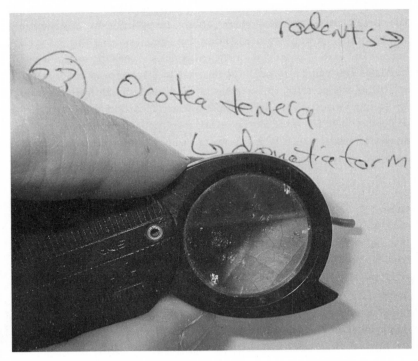

FIGURE 10.12. This wild avocado tree species, *Ocotea tenera*, will attract spectacular birds—like the resplendent quetzal and the three-wattled bellbird—with its
fruit if it survives to adulthood. The leaves of this tree, even at the sapling stage,
have tiny domicia—small houses—where arthropods can live. Photograph by
Eben Kirksey.

FIGURE 10.13. Milton planted bitterbush (*Picramnia antidesma*), a common shrub found throughout the lowlands of Central America and the Caribbean, in Monteverde's regenerating cloud forest to test the boundaries of its range amid changing environmental conditions. The bark of this bush "has been given with success as an alternative in constitutional affections, connected with syphilis and yaws, and as a tonic in debility of the digestive organs, and in intermittent fever," according to James Macfadyen's *The Flora of Jamaica*, published in 1837. Photograph by Eben Kirksey.

savanna oak (*Tabebuia rosea*, Bignoniaceae)—to the Monteverde Cloud Forest School. The company was planting fifteen thousand trees in the region to protect their hydroelectric projects from erosion. One of the three ICE trees, the cocobolo, has fine wood used in furniture and jewelry. It is currently listed as "vulnerable" on the IUCN Red List of Endangered Species. Wood of the other two trees, the cigar-box cedar and the false savanna oak, is often used by Costa Ricans for building houses. Milton later told us that all of the cocobolos, the cigar-box cedars, and the false savanna oaks we had planted in 2008 had died within the first year. Bitterbush was the only lowland plant that Eladio and I found growing at the Monteverde Cloud Forest School during our 2014 survey.

Many expatriate biologists in Monteverde fear that attempts to prepare for climate change may backfire. Human actions often trigger unexpected outcomes. Introducing lowland species to the highlands, according to local critics, might hasten the demise of members of local flora and fauna that are struggling to survive. Milton disagrees. Rather than seeing broken ecosystems, with species being popped out like rivets in an airplane wing, he is helping fasten an emergent community in place—testing out new living rivets all the time, to see if they might fit. By his reckoning, planting lowland trees in Monteverde will ensure that a forest will still be standing even if local trees die from heat stress.

William Haber, who has been studying the ecology of Monteverde's cloud forest since 1973, is not worried that reforestation projects like Milton's somehow violate the natural order (as many expatriate critics imply). "To me it doesn't make too much difference if you just let it regenerate," Haber told me, "or if you plant things that you like for a specific reason because you have a goal in mind." Haber arguably knows more about the distribution and diversity of Monteverde's trees than anyone else alive today. "The Plants of Monteverde," the results of Haber's collaboration with the Missouri Botanical Garden and the Electronic Field Guide Project at the University of Boston, currently serves as the definitive reference for anyone conducting ecological research in the region.[59] Pointing me to the local paleontological record in Monteverde, Haber related his bleak visions of a future when a geological cataclysm will eclipse all local human agency and action.

Haber situates local ecologies within a long regional history of volcanic disasters. Ash, pumice, and cinders from Arenal Volcano have periodically destroyed the forests of Monteverde, with eruptions taking place about every five hundred years.[60] "The last big ash layer, the last major explosion, was about a thousand years ago," Haber told me. "We are about due for another big one." These prehistoric disasters likely had dramatic impacts on the forests of Monteverde. "I presume a lot of it was killed and then it was recolonized," Haber said. "So sometimes when you think you are in primary forest, it is actually still a regenerating forest." A major volcanic eruption at Arenal would certainly kill much of the local flora and fauna of Monteverde and also disrupt long-term conservation agendas. But against the backdrop of this potentially bleak future, amid the immanent possibility of an apocalyptic ending, there are still signs of hope.

Volcanic ash fields are fertile grounds for nourishing emergent ecological assemblages or agricultural initiatives. A team of archaeologists has found that the village lifestyle of indigenous groups "was remarkably stable and resilient in spite of the effects of at least nine prehistoric explosive eruptions of Arenal Volcano." By comparison, other prehistoric sites near volcanoes—in Mesoamerica, the Andes, and Middle America—had much more dramatic disruptions following eruptions. In contrast to these

FIGURE 10.14. Evidence of past volcanic disasters is clearly visible in sections of the soil that have been cut away during the construction of roads in Monteverde. Photograph by Eben Kirksey.

other sites, which showed evidence of political stratification, economic centralization, and reliance on commodities transported great distances, the prehistoric villages in the shadows of Arenal were sparsely populated and avoided reliance on irrigation systems, storage facilities, and large-scale construction. Archaeologists attribute the "striking stability" of prehistoric societies in the greater Monteverde region to "a diversified adaptation within the environment." Wild food sources—fish from a nearby lake, game animals from the forest, fruit trees, root crops, wild seeds, and berries—were supplemented with agriculture focused on corn and other seed crops.[61] In other words, practices of gardening in ruins have a deeply rooted local history.[62]

Looking past many potential catastrophes looming on future horizons—predicted extinction events, possible volcanic eruptions, and reports of economic disaster in distant lands—Milton sees the possibility of fresh order-forming associations amid order-destroying disasters. Milton has seeded a forest that will endure many different possible futures. By paying careful attention to the interests of entrepreneurial agents, he has brought a swarming multitude to life.[63] Rather than envisioning a postapocalyptic utopia, where humans have abruptly left the scene, Milton imagines a shared future with plants and animals he loves.[64] Instead of holding loved ones in a tight, uncomfortable embrace, he has tried to keep a polite distance—experimenting with forms of care and generative practices of disinterest. Dismissing the immediate relevance of apocalyptic predictions, and people who are waiting for the arrival of messianic figures, Milton has grounded modest hopes by gardening in twentieth-century ruins. His visions of the future are full of happiness. Living with the hap of what happens, multiple species are flourishing as they navigate the contingency of a world managed with tactful oversight.[65]

CONCLUSION

. . .

Ecosystems have become assemblages that increasingly cannot be separated from human social systems and technical machines.[1] The Panamanian forest plus the canal system becomes a watershed, a natural infrastructure that will endure development plans of the future. A wetland plus a fangueo tractor plus the transnational flow of capital becomes a duck factory. A glass aquarium plus mutant fruit flies plus an endangered frog becomes part of the Amphibian Ark. By offering thick descriptions of ecosystem parts, which also serve as partial sketches of shifting and ephemeral wholes, my intention has been to account for emergent conditions of entangled coexistence. As institutions create and sustain hybrid ecosystems, integrating machine parts and biotic elements into contingent associations, a riotous diversity of life is also going wild beyond the reach of human dreams and schemes.

Emergent ecologies are being shaped by a multitude of creative agents. Multispecies communities live in the minds of creatures like *Ectatomma* ants, conscious animals who actively care for other beings and things. Convivial plants, like *Inga punctata*, also form assemblages with nectaries on their stems, sugary treats in their fruits, and nutritious exudates from their roots. Rather than just building niches for themselves—constructing and altering environments for members of their own kind—enterprising critters like earthworms and their diverse companions in the rhizosphere integrate the interests of others as they bring ecosystems together.[2] Flourishing in the aftermath of order-destroying disruptions—in abandoned cattle pastures, in the detritus of industrial food production, after disease outbreaks, and geological disasters that eclipse human agency—these enterprising critters are opening up the future to lively possibilities.

Early theorists supposed that ecological communities would predictably reemerge in the same form if destroyed, while later conservationists worried that they might fall apart if important rivets were popped out with species extinctions. Rather than collapsing systems, I found unruly assemblages flourishing and proliferating in unexpected places. Rather than natural units of vegetation reemerging in a predictable process of succession, I found novel ecosystems fastened into place with new rivets and cyborg articulations. Accepting that ecological communities are dynamic, ever-changing systems—with parts that can be taken away or added—opens up ethical and practical dilemmas. One question, first posed by Matthew Chrulew, should be asked again and again: "How should we love in a time of extinction?"[3] Loving frogs that have been orphaned from their original ecosystems might mean capturing them—subjecting them to "the ambiguous grace of salvation."[4] Love for others who are flourishing in an era of extinction, like the promiscuous *Ectatomma* ants of Central America or the wild rhesus macaques of Florida, might best be consummated with coy invitations to fleeting encounters that always leave open the possibility of escape. Edible companion species—like rice, cattle, or ducks that we care for only to kill—sometimes impinge on others whom we love. Our affection for these species is opening up both opportunities and obstacles for the organisms concerned as well as the multispecies communities where they live.

This key question of emergent ecologies—how should we love?—often comes with preordained answers in line with dominant political projects and economic interests. Political ideas about belonging offer deceptively simple plans of action: kill the aliens, love the natives. Chemical industries, with vested economic interests in monocrop agriculture, offer answers based on the same kind of deadly distinctions: poison the pests, fertilize the crops. Conservation practitioners, policy makers, and farmers are already moving beyond these seemingly simple distinctions. People belong to heterogeneous political communities and are starting to move beyond xenophobic discourses of the twentieth century. Restoration ecologists and nature lovers are finding that wayward and unpredictable forms of life can generate new niches, habitat, and opportunities for symbiotic attachments even as they disrupt established assemblages.

Killing often becomes necessary when one makes commitments to others within ecological communities. Taking a stand for creatures one loves means taking a stand against enemies that present existential threats to them. If you are caring for frogs in your own home, and your artificial ecosystem becomes infected with *Bd*, killing chytrids is relatively simple: bleach the tanks and bathe frogs in an antifungal betadine treatment. If you care for the diverse plants and animals living in Costa Rican cloud forests and would like to welcome them back to abandoned pastures, a logistically difficult challenge is ahead: killing star grass. Figuring out how to kill is just as important as figuring out whom to kill. Cattle can certainly kill and control star grass; they can twist death back into life. Convivial microbes might one day help endangered frogs kill chytrids, allowing them to live outside artificial ecosystems. Final solutions, projects aimed at eradicating all perceived enemies, rarely work—for practical reasons, as much as political and ethical reasons. It is probably not possible to completely eliminate African grasses from Central America, or pathogenic chytrids from waterways, or Burmese pythons from Florida, no matter how hard one tries. Love in ecological communities means living with the necessary labor of killing. It also means consciously dwelling with the genuine ethical difficulties that accompany attempts to promote good lifeways.[5]

When we participate in market economies, or grow our own vegetables, we are casting our lots with some ways of life and not others.[6] Life and death are at stake every time we eat, buy clothes at the store, or flip an electric light switch. Unescapable entanglements with powerful assemblages that telegraphically mete out bad deaths and cascading chains of destruction in ecological communities have led many environmental advocates to sadness and cynicism. Despite a groundswell of popular action, global climate change is outpacing all attempted solutions. Capitalist enterprises that are rapidly destroying forests and wetlands in diverse corners of the globe may well be unstoppable. Recognizing these new facts of life has led many to feelings of futility. Critics hover around the halls of the academy, lambasting all for complicity in these assemblages, but doing little to love, live, and fight. Alongside legions of cynical critics, as well as communities of experts and working professionals promoting corporate and government interests, a multitude of tinkerers and thinkers are transforming feelings of futility into concrete action, cynicism into happiness and hope.[7]

This multitude is forging convivial alliances, exploring wild possibilities. Sustaining multispecies families, tending compost bins, and sorting through the detritus of capitalism, these creative agents are caring for things and beings that are ready at hand. Organic intellectuals are gardening in the ruins, finding delectable fruits and mushrooms in blasted landscapes, finding new companions while wandering through haunted and weedy wrecks.[8] Microbes are awakening pleasures in us that eclipse purely utilitarian calculations; animals and plants are stirring our interests and desires, using us to multiply their forces.[9] What sorts of novel ecological assemblages might we build together? Can we construct new ecosystems while embracing social justice concerns, grappling with the subjective experiences of other organisms, and upholding conservation values all at the same time? Can we craft tactful proposals to those whom we love, offering links to our social worlds and industrial supply chains while keeping windows open that give them opportunities to escape?

ACKNOWLEDGMENTS

The origins of this book can be traced to my time as an undergraduate at New College of Florida, when Sandra Gilchrist, Alfred Beulig, and Maria Vesperi gave me a long leash and guided me as I worked to integrate knowledge of biology and cultural anthropology on my own terms. I first traveled to Costa Rica in 1996, when I served as an undergraduate research assistant for Terry McGlynn at La Selva Biological Station. Staff scientists at the Smithsonian Tropical Research Institute—namely William Wcislo, Allen Herre, Egbert Leigh, and Stanley Rand—first welcomed me to their facilities as an undergraduate in 1997. When I returned a decade later, as a National Science Foundation Postdoctoral Fellow in Science and Society (award number 750722), Smithsonian staff graciously let me explore questions related to ecological entanglements, cultural theory, and interspecies sociality while in residence at their laboratories.

James Clifford, Donna Haraway, and Anna Tsing, who gave me a set of critical tools for thinking about nature-culture, contact zones, and multispecies mingling, allowed me to take leave from my PhD program at UC Santa Cruz in 2006 to pursue graduate training in field ecology. The Organization of Tropical Studies graciously offered me a spot in their fundamentals graduate course, Tropical Biology: An Ecological Approach. Gregory Gilbert helped me reactivate my knowledge of tropical ecology at UC Santa Cruz and also provided financial support as I took the course. Erika Deinert, Katja Poveda, and Diego Salazar taught me how to dance merengue while giving me rigorous training in the methods of field ecology. Don Carlos Porras provided critical logistical support, while Craig Guyer taught me how to speak with and for fringe-toed foam frogs. Peers in this tropical biology course—especially Quinn Long, Katie Cramer, Elizabeth Wheat, Haldre Rogers, Laura Marx, Amy Savage, and Kristiina Hurme—helped me understand emergent ecological thinking.

Geoffrey Bowker and Susan Leigh Star helped me find my feet as a freshly minted PhD, serving as committed guides as I began wading through swamps and wandering through tropical forests throughout the Americas. I am deeply indebted to the generosity of many people who are quoted in this book—especially Heidi Ross, Edgardo Griffith, Brian Gratwicke, Angie Estrada, Jayne Reyes, Joyce Longcore, Bob Gottschalk, Sally Lieb, Tish Hennessey, Jenny Novak, Wendy Adams King, and Milton Brenes. The Cruz-Obando family—Eladio, Anais, Arturo, Jose Andreas, Ana Rita, and Kim, as well as little Randy and Cassidy—kept me supplied with fresh coffee and gallo pinto as I labored on my past book projects. Jesse Griest and the staff at Centro de Educación Creativa—Sara Haughton, Scott Timm, and Cristina Castro in particular—helped me feel at home in the Monteverde community. As I transitioned to Costa Rica's lowlands, taking up residence at the OTS Palo Verde Biological Field Station in Guanacaste, Mahmood Sasa showed me how to find humor and hope in a blasted landscape. Gilberto Murillo, Romelio Campos, Juan Serrano, and Pajarin gave me crucial logistical support. Daniela Marini assisted me with fieldwork in Palo Verde and helped me understand the precarious state of life in Bagatzí while we worked together in the United States.

Maria Vesperi and Jay Sokolovsky provided me with a home base as I started tracing tales of the mystery monkey and studying the emergent ecologies of Florida. Jono Miller and Julie Morris pointed me toward surprising lines of flight and also saved me from the blood-sucking Siphonapterans. Teaching Ethnographies of Nature to talented students at New College in 2008—especially James Birmingham, Nancy Rose Spector, Brooke Denmark, and Sam Chillaron—helped me animate some of my latent knowledge of nature and culture in Florida. Erin Riley, Tiffany Wade, Elan Abrell, and Amanda Concha-Holmes joined me for collaborative field research on the Silver River in 2013. Scott Mitchell graciously opened up access to the Silver River Museum and cabins in the Ocala National Forest, while Sally Lieb gave the research team open access to the park. Alice Bard of Florida's Department of Environmental Protection granted us research permit number 1206113 to conduct research on the wild rhesus monkeys in Silver Springs State Park. Anderson Flack was a diligent research assistant during our hunt for wild monkeys on the Ocklawaha River.

Louise Lennihan has been a steadfast supporter and backer of this project—encouraging me to wander the world as a postdoctoral fellow at the CUNY Graduate Center from 2010 to 2012, and then giving me the keys back to my old office in 2014 as I put the finishing touches on the manuscript. The Committee for Interdisciplinary Science Studies at CUNY provided me with intellectual nourishment and stimulation. Jesse Prinz, Joan Richardson, Victoria Pitts-Taylor, Erin Glass, Murphy Halliburton, Ashley Dawson, Vincent Crapanzano, Betsy Wissinger, Abou Ali Farman Farmaian, Peter Godfrey Smith, and Joe Alpar were among my smartest and most generous CUNY colleagues. Ray Ring, Reggie Lucas, and Chris Lowery provided critical support behind the scenes. Members of the NYC multispecies writing group—Paul Kockelman, Astrid Schrader, Patricia Clough, Amy Herzog, Traci Warkentin, and Jeffrey Bussolini—helped me hone the ideas in this book. Colleagues in wider New York City networks—namely Jackie Brookner, Orit Halpern, Lori Gruen, Lee Quinby, and Paige West—also provided critical support. Isabelle Stengers, Natasha Myers, Matei Candea, Stefan Helmreich, Heather Paxson, Steven Meyer, Dorion Sagan, and Cary Wolfe stimulated my thinking about emergent ecologies during their presentations at the CUNY Cosmopolitics Symposium. Students in my Ethnography of Science course at CUNY, especially Elan Abrell and Charlie Lotterman, helped me think through our joint amphibious adventures.

Tammy Pittman and Krista Dragomer gave me an experimental arena at Proteus Gowanus for testing out ideas in formation about performativity and science. Collaborators who took part in the Xenopus Pregnancy Test—Charlie Lotterman, Dehlia Hannah, and Lisa Jean Moore—helped me tinker with and rethink these ideas. Members of the Treehaus—a Brooklyn collective of artists, musicians, activists, and scholars—let me hack into an unwanted refrigerator and stage performative experiments in their living room. Grayson Earle was a key collaborator who coproduced artworks with me and helped stage underground frog pregnancy tests at multiple undisclosed locations in Brooklyn. Mike Khadavi built a living ecosystem inside *The Utopia for the Golden Frog* and opened doors into frog worlds.

Joining the Environmental Humanities program at UNSW gave me the time and institutional support necessary to finish this book. Thom van Dooren, a master at niche creation, generated happy conditions of possibility for me in Sydney. The broader Environmental Humanities community formed by Deborah Bird Rose, Stephen Muecke, Emily O'Gorman, Jennifer Hamilton, Hollis Taylor, Kate Wright, Matthew Kearnes, Karin Bolender, Laura McLauchlan, and many others provided me with intellectual stimulation as I brought this project to fruition. Vanessa Lemm helped me understand the Nietzschean genealogy of emergence. Sally

Pearson, Taline Tabakyan-Golino, Anne Engel, and James Donald also provided critical support.

Thanks are also due to Kathy High, Adam Zaretsky, Robert Nideffer, Jane Kang, Susan Harding, Gina Clark, Ruth Wallen, Eduardo Kohn, Beatrice Marovich, Raymond Corbey, Annette Lanjouw, Jon Stryker, John Hartigan, Kathleen Stewart, Irus Braverman, James Scott, Michael Marder, Celia Lowe, Emma Kowal, Joanna Radin, Kim and Mike Fortun, Brandon Costelloe-Kuehn, Daniel Münster, Ursula Münster, Anne Spalliero, Carla Freccero, Beatriz da Costa, Bettina Stoetzer, Eva Hayward, Elaine Gan, Joe Dumit, Tim Choy, Rich Pell, Rich Scaglion, Lindsay Kelley, Natalie Loveless, Sandra Koelle, Scout Calvert, Rebecca Schein, Becca Munro, Tilly Hinton, Ryan the Fish, Steve the Frog, Steve the Human, Loretta the Frog, and the rest of the Treehaus crew. Heroic work by Santiago Meneses, involving laboratory labor with *Ectatomma* ants late into the night, is enabling me to translate some of the insights of this book into the language of the biological sciences. Presenting chapters of this monograph at intellectual centers around the world has greatly improved the project: Cornell University, SUNY Buffalo, UT Austin, Rice University, Rensselaer Polytechnic Institute, Yale University, Macquarie University, the University of Sydney, the University of Technology Sydney, Goddard College, Kyoto University, the Rachel Carson Center, the Arcus Foundation, Heidelberg University, Leiden University, Columbia University, University of South Florida, Carnegie Mellon University, University of Pittsburgh, and University of Washington. Panels at the American Anthropological Association, the Society for Social Studies of Science, and the American Association for the Advancement of Science also helped me refine my thinking.

Sarah Franklin deserves special thanks for her role as a smart and generous peer reviewer of my last two book projects. Agustín Fuentes offered early advice as I began exploring multispecies contact zones and made the book sing with his useful review. My parents, Jane and Will Kirksey, both gave me love and logistical support as I chased after elusive research questions in far-off lands. My dad read the final manuscript as it was written, offering encouragement and copyediting in real time.

Ken Wissoker has been a tactful guide, a sage mentor, and a consistent cheerleader over the years as I wove diverse threads into a unified monograph. Jade Brooks has been a speedy shepherd, nudging me and my peer reviewers along to keep on a very efficient publication timetable. Thanks are due to the extended team and alumni at the Press—especially

Laura Sell, Sara Leone, Julie Thomson, Christine Riggio, Bonnie Perkel, Amy Ruth Buchanan, Katie Courtland, Helena Knox, Chad Royal, and Tim Elfenbein—for past, present, and future help. The careful eyes and steady hands of Karen M. Fisher and Liz Smith turned this manuscript into a book. Elan Abrell burned the midnight oil to whip the notes, bibliography, and photographs into tip-top shape.

This book is dedicated to my sister and best friend, Kate Kirksey.

NOTES

INTRODUCTION

1 Lowe, "Viral Clouds."
2 Kirksey, Shapiro, and Brodine, "Hope in Blasted Landscapes."
3 Emergence equates with *entstehung*, the moment of arising, in a classic essay: Foucault, "Nietzsche, Genealogy, History," 148–52.
4 The Clements-Gleason controversy has itself been the subject of much historiographical debate. Geoffrey Bowker recounts the debate to suggest ecosystems belong to a "class of things which it is hard to classify" (*Memory Practices in the Sciences*, 144–45). Comparing their work to the plotlines of modernist and postmodernist fiction, Debra Journet suggests, "Although Gleason's and Clements's theories were not necessarily influenced by fiction, they parallel it in remarkable ways. . . . Notions of coherence, direction, and progress—or of their opposites—become part of the way groups of people make sense of reality" ("Ecological Theories as Cultural Narratives," 467). Christopher Eliot, a philosopher, suggests that the difference between Clements and Gleason lies "in methodology, not metaphysics" ("Method and Metaphysics in Clements's and Gleason's Ecological Explanations," 86). Clements, *Research Methods in Ecology*, 282, 302; Clements, *Plant Succession*, 30; Clements, "Nature and the Structure of the Climax," 254; Gleason, "The Individualistic Concept of Plant Association," 16, 7.
5 The lasting value of Tansley's 1935 paper "comes not in his concluding litany of terminology (most of which is now altered or abandoned)," conclude Joe Mascaro and colleagues, "but in his flashes of uncertainty about nature" ("Origins of the Novel Ecosystems Concept," 45–47). These uncertainties relate to the instability of the "biotic communities" described by earlier researchers like Clements. "Many ecologists hold that all vegetation is always changing," to quote from Tansley's original 1935 paper. "It may be so: we do not know enough either to affirm or to deny so sweeping a statement." Tansley also understood that the nonbiotic environment, the physical factors controlling the climate, were also in flux. Interestingly, Tansley

regarded ecosystems themselves as artifacts of analysis. Tansley, "The Use and Abuse of Vegetational Concepts and Terms," 299–301.

6 Ehrlich and Ehrlich, *Extinction*, xi.

7 Mascaro, "Perspective," 156.

8 Yung et al., "Engaging the Public in Novel Ecosystems," 248.

9 Standish et al., "Concerns about Novel Ecosystems," 298.

10 Mascaro et al., "Origins of the Novel Ecosystems Concept," 46.

11 Three parallel dynamics take place in situations of cultural contact—a process of losing the previous culture (deculturation), the acquiring of another culture (acculturation), and the creation of new cultural phenomena (neoculturation)—according to Fernando Ortiz, an early twentieth-century Cuban anthropologist who studied colonial encounters. Ortiz's 1940 book, *Cuban Counterpoint*, anticipated contemporary scholarship by studying how biological agents influence cultural dynamics. He chronicled how two plants—tobacco and sugar—captured the imagination of Europeans and American Indians, prompting diverse societies to work together in creating novel economic, political, and cultural systems. Ortiz, *Cuban Counterpoint*, 102.

12 Cyborgs also "take pleasure in the confusion of boundaries" and make arguments "for responsibility in their construction." Haraway, "Manifesto for Cyborgs," 65.

13 Stengers, *Cosmopolitics I*, 35–36.

14 Stengers, *Cosmopolitics I*, 35–36. See also Latour, *Politics of Nature*, 455.

15 Margulis and Sagan, *Acquiring Genomes*, 20.

16 "The 'symbiotic agreement' is an event," writes Stengers, "the production of new, immanent modes of existence, and not the recognition of a more powerful interest before which divergent particular interests would have to bow down." Stengers, *Cosmopolitics I*, 35.

17 Stengers, *Cosmopolitics I*, 35–36.

18 Amin and Thrift, *Arts of the Political*, 132.

19 Wolfe, "Introduction," xiii.

20 Serres, *The Parasite*, 253.

21 Janzen, "Herbivores and the Number of Tree Species in Tropical Forests"; Connell, "On the Role of Natural Enemies."

22 Paul Kockelman has written a brilliant account of enemies, parasites, and noise—critically reviewing Michel Serres's work and bringing it into conversation with Peirce's notion of thirdness. He notes that Serres spends very little time on interpretation (or code), and instead focuses his efforts on circulation (or channel). Formulating his own technical definition, Kockelman asserts that the parasite inhabits the implications of this statement: "An object (action or sign) considered as a means to an end (or infrastructure considered as a path to a destination) is a second (or intermediary), but insofar as it implies (embodies or indexes) other ends it might be diverted to serve, or indeed implies any way it may fail to serve an end

(whether original or diverted), it is a third (or mediator)." Kockelman, "Enemies, Parasites, and Noise," 412–13; see also Serres, *The Parasite*.

23 Extinction studies is a rapidly growing field. Genese Sodikoff has edited a key point of reference for anthropology: *The Anthropology of Extinction*. My colleagues in the Extinction Studies Working Group—in particular Matthew Chrulew, Deborah Bird Rose, Thom van Dooren, and Michelle Bastian—have deeply influenced my thoughts on the subject. For their latest publications on the subject, see Extinction Studies Working Group, http://extinctionstudies.org.

24 Compare this notion of the "ontological amphibian" with ten Bos, "Towards an Amphibious Anthropology," 74; Sloterdijk, *Bubbles*.

25 The swarm has generated mutations in forms of association that are "structurally innovative but politically ambivalent," according to Eugene Thacker. Pentagon strategists have begun to imagine how soldiers might become animals, deploying the tactics of swarming on the front lines of the global war on terror. Still, many artists and allied intellectuals are playfully reappropriating these tactics of war. Thacker, "Networks, Swarms, Multitudes"; Kosek, "Ecologies of Empire"; Kirksey et al., "Poaching at the Multispecies Salon"; Hardt and Negri, *Multitude*.

26 Vivanco, "The Work of Environmentalism," 132.

27 Taussig, *Shamanism, Colonialism, and the Wild Man*, 212; Franklin, "Ethical Biocapital," 102.

28 Vivanco, "The Work of Environmentalism," 132.

29 Haraway, *When Species Meet*, 295; Puig de la Bellacasa, "'Nothing Comes without Its World,'" 198.

30 Timothy Ingold exemplifies the very best of this work in this tradition. See Ingold, *The Perception of the Environment*.

31 Many philosophers, such as Matthew Calarco, assert that "the human-animal distinction can no longer and ought no longer be maintained" (*Zoographies*, 3). But such a statement naturalizes the very categories that must be troubled. Shifting to the idiom of multispecies ontologies offers a way to radically rethink binaries like human-animal, plant-fungus. Derrida, *The Animal That Therefore I Am*.

32 Marder, *Plant Thinking*, 85.

33 Kohn, *How Forests Think*, 72.

34 For more on the "contingencies of unexpected connections," see Tsing and Pollman, "Global Futures," 107. For more on the methods and tactics of multispecies ethnography, see Kirksey and Helmreich, "The Emergence of Multispecies Ethnography"; Kirksey, *The Multispecies Salon*.

35 Kirksey, Schuetze, and Helmreich, "Introduction," 9; da Costa and Philip, *Tactical Biopolitics*.

36 See Lestel, Brunois, and Gaunet, "Etho-ethnology and Ethno-ethology"; Hodgetts and Lorimer, "Methodologies for Animals' Geographies," 3.

37 Berlant, *Cruel Optimism*, 24, 64; Choy, *Ecologies of Comparison*, 28; Ahmed, *The Promise of Happiness*, 31.

38 See Jameson, "Marx's Purloined Letter," 63–64.

39 Kirksey, Shapiro, and Brodine, "Hope in Blasted Landscapes," 39; Tsing, "Blasted Landscapes"; Crapanzano, *Imaginative Horizons*.

40 Haraway, "Speculative Fabulations for Technoculture's Generations," 253.

ONE. PARALLAX

1 Biff Bermingham, "Message from the Director: December 20, 2010," Smithsonian Tropical Research Institute, http://www.stri.si.edu.

2 Leigh, *Tropical Forest Ecology*, 12.

3 Strain, "Stereoscopic Visions," 51.

4 Compare Strain, "Stereoscopic Visions," 47; Tilley, *Africa as a Living Laboratory*.

5 Leigh, *Tropical Forest Ecology*, 3.

6 Clipping from *Science*, June 13, 1924, 521, RU 134, Box 1, Folder 4 of 12, Smithsonian Institution Archives, Washington, DC.

7 "The Tropical Biological Station on Barro Colorado Island in the Panama Canal Zone," n.d., RU 134, Box 1, Folder 8 of 12, Smithsonian Institution Archives, Washington, DC.

8 Carse, "Nature as Infrastructure," 547.

9 James Zetek, "The Barro Colorado Island Biological Laboratory," Panama, August 1931, pp. 12–13, RU 134, Box 12, Folder 1 of 10, Smithsonian Institution Archives, Washington, DC.

10 James Zetek, "Third Annual Report of the Barro Colorado Island Research Station," March 1, 1927, RU 134, Box 1, Folder 7 of 12, Smithsonian Institution Archives, Washington, DC.

11 Henson, "Invading Arcadia," 577.

12 Rose and van Dooren, "Unloved Others."

13 "Architectures of apartheid," in the words of Geoffrey Bowker and Susan Leigh Star, involve "a mixture of brute power, confused eugenics, and appropriations of anthropological theories of race." Bowker and Star, *Sorting Things Out*, 28.

14 Henson, "Invading Arcadia," 587.

15 Zetek to Barbour, May 5, 1927, RU 134, Box 1, Folder 12 of 12, Smithsonian Institution Archives, Washington, DC.

16 "BCI was triply isolated," in the words of Catherine Christen, "as an island, as a U.S. territory, and as a scientific reserve." Christen, "At Home in the Field," 541.

17 Ventocilla and Dillon, *Gamboa*, 25, 124.

18 Harvey, *Spaces of Hope*, 161–63.

19 Accessed April 27, 2015, https://www.flickr.com/photos/killbyte/4613230214.

20 Carse, "Nature as Infrastructure," 549, 554–55 (citing Wadsworth).

21 Leigh, *Tropical Forest Ecology*, 11.

22 The UN estimated 500 deaths; the Commission for the Defense of Human Rights in Central America estimated 2,500–3,000 deaths; and the Commission for the Defense of Human Rights in Panama (Comisión Nacional de Derechos Humanos de Panamá) estimated 3,500 deaths. For further background, see Lindsay-Poland, *Emperors in the Jungle*, 118–23.

23 Cosmopolitans, as described by Immanuel Kant, are dangerous nomads; in the eyes of Isabelle Stengers, they are irredeemably destructive or tolerant without attachments to common worlds. Stengers, "The Cosmopolitical Proposal."

24 Following the introduction of the electric ant to the Galapagos Islands, it drastically reduced and even extinguished the populations of every other ant species it encountered, including two endemics found nowhere else on the planet. Clark et al., "The Tramp Ant *Wasmannia auropunctata*."

25 Balakrishnan and Schmitt, *The Enemy*, 108.

26 A series of classic papers about these dynamics in the 1990s by Michael Breed, a senior behavioral biologist, concluded that "thievery" was taking place among *Ectatomma* ants. Breed suggested that individual "thief ants" use chemical camouflage to gain access to neighboring colonies. "Thief ants have reduced quantities of cuticular hydrocarbons on their surface," Breed reported, "and their cuticular hydrocarbon profile is intermediate between the hydrocarbon profile of their own colony and the colony from which they are stealing." While conducting my own research as an undergraduate on Barro Colorado Island in 1997, using the methods of behavioral ecology, I began to suspect that there was more to the story than "stealing." Starting with more generous assumptions about the nature of others, I began to investigate the possibility that *Ectatomma* ants were "trading." Breed et al., "Thief Workers and Variation in Nestmate Recognition," 327; see also Breed et al., "Thievery, Home Ranges, and Nestmate Recognition"; Breed et al., "Acquired Chemical Camouflage in a Tropical Ant."

27 These findings were highly statistically significant. As an undergraduate, I conducted a series of experimental introductions from February through June 1997 on Gigante Peninsula in the Barro Colorado Nature Monument, Panama, and in June 1997 at La Selva Biological Station in Costa Rica. A total of 953 ants from 94 different nests were used in 71 experimental trials. Running statistical tests on these data, I found highly significant differences in how ants from focal home nests were treated with respect to ants from neighboring nests or far away ($p < 0.001$, three-way interaction). I also found highly significant paired interactions for all three groupings: home versus far ($p < 0.001$), near versus far ($p < 0.001$), and home versus near ($p < 0.01$). While I was observing these nests, I also recorded the numbers of unmarked ants (not manipulated during the experimental trial) that were dragged out of the nest by guards. More unmarked ants were dragged

away (67 in total) than the combined total of ants I introduced from the home, near, and far nests. Drawing on these data, I submitted a manuscript, titled "Fuzzy Boundaries: Inter-nest Exchange and Levels of Conspecific Recognition in the Ground Ant, *Ectatomma ruidum*," to the journal *Animal Behavior* in 1998. One reviewer, in recommending the manuscript's rejection, wrote, "The study is not of sufficiently broad interest to appeal to a diverse audience."

28 In the late 1970s, Wilson defined sociobiology as "the extension of population biology and evolutionary theory to social organization" (*On Human Nature*, xx). Stephen Jay Gould, one of Wilson's colleagues at Harvard, was among the key early critics of his ideas within the field of biology— contending that reductive explanations linking genes to human behavior (like aggression) failed to account for social and ecological influences on behavior (Gould, *The Mismeasure of Man*). While I was conducting this research in Panama, working with people who described their own work in terms of "behavioral ecology," Wilson's ideas still held much sway in the community of ant experts.

29 There is considerable deviation from this ideal type. In many ant colonies, those of *Ectatomma* included, multiple queens can be found inside. Workers also lay eggs—some that are eaten by other adults and others that develop into larvae. Male ants—with wings, small heads, and a waspy look—take little part in colony life other than mating. Hora et al., "Facultative Polygyny in *Ectatomma tuberculatum*"; Hora et al., "Egg Cannibalism in a Facultative Polygynous Ant"; Shik and Kaspari, "Lifespan in Male Ants Linked to Mating Syndrome"; Lachaud et al., "Queen Dimorphism and Reproductive Capacity."

30 Hölldobler and Wilson, *The Ants*, 179.

31 Hardt and Negri, *Multitude*, 56–57.

TWO. ONTOLOGICAL AMPHIBIANS

1 Ten Bos, "Towards an Amphibious Anthropology," 74; Sloterdijk, *Bubbles*.

2 Donna Haraway's work has also deeply influenced me here. "If we appreciate the foolishness of human exceptionalism," she writes, "then we know that becoming is always becoming *with*—in a contact zone where the outcome, where who is in the world, is at stake." Haraway, *When Species Meet*, 244.

3 Here I am engaging in the practice of poaching—converting Sloterdijk's texts through reading like a trespasser on a private hunting reserve of the elite literati. If chefs "poach" pears, using red wine and honey to intensify and transform the flavor of the fruit, in poaching "ontological amphibians" my wine is Stengers's notion of cosmopolitics and my honey is the parasite of Serres. See de Certeau, *The Practice of Everyday Life*, 171; Matsutake Worlds Research Group, "Thoughts for a World of Poaching"; Kirksey, *The Multispecies Salon*.

4 Stengers, *Cosmopolitics II*, Book 7.

5 Stengers, "The Cosmopolitical Proposal," 995.

6 For a discussion of the enemy-ally distinction, in the context of Carl Schmitt's work on the subject, see Kirksey, *Freedom in Entangled Worlds*, 230n9.

7 Latour, *Politics of Nature*, 455.

8 Cook, "Cotton Culture in Guatemala," 485.

9 In contemporary German, the word *umwelt* simply means "environment." During a July 2014 visit to the University of Heidelberg, I discovered a bilingual sign in my hotel asking me to think of the environment/umwelt by reusing my towel. In von Uexküll's work, and the subsequent work in Continental philosophy, umwelt signifies something more specific: the species-specific phenomenological world inhabited by all animals and possibly other organisms like plants, fungi, and microbes.

10 Ten Bos, "Towards an Amphibious Anthropology," 74.

11 Von Uexküll, "A Stroll through the Worlds of Animals and Men," 323.

12 Giorgio Agamben's careful reading demonstrates that von Uexküll did not privilege human perceptions of reality:

> Uexküll begins by carefully distinguishing the *Umgebung*, the objective space in which we see a living being moving, from the *Umwelt*, the environment-world that is constituted by a more or less broad series of elements that he calls "carriers of significance" (*Bedeutungstrager*) or of "marks" (*Merkmaltrager*), which are the only things that interest the animal. In reality, the *Umgebung* is our own *Umwelt*, to which Uexküll does not attribute any particular privilege and which, as such, can also vary according to the point of view from which we observe it. There does not exist a forest as an objectively fixed environment: there exists a forest-for-the-park-ranger, a forest-for-the-hunter, a forest-for-the-botanist, a forest-for-the-wayfarer, a forest-for-the-nature-lover, a forest-for-the-carpenter, and finally a fable forest in which Little Red Riding Hood loses her way.

> Agamben, *The Open*, 40–41; Von Uexküll, *A Foray into the Worlds of Animals and Humans*, 42.

13 Ingold, *The Perception of the Environment*, 176.

14 Agamben, *The Open*, 45.

15 Since von Uexküll's work on ticks in the early twentieth century, biologists have discovered that they can attend to other stimuli, like "assembly pheromones" that are secreted by an individual while sucking blood, drawing their extended kin to a food source. James Hatley has taken these conversations in a different direction. Writing of how the emerging world of the Anthropocene is reverberating into the umwelt of ticks, Hatley asks, "How should we live our relationship with this co-evolved and living world? How are we called to witness a more-than-human living world?" Hatley, "Blood Intimacies and Biodicy"; Oliver, "Biology and Systematics of Ticks," 418–19.

16 Von Uexküll, "A Stroll through the Worlds of Animals and Men," 325.

17 Buchanan, *Onto-Ethologies*, 93, 189.

18 Quoted in Agamben, *The Open*, 51. See also Santer, *On Creaturely Life*, 8.

19 Sloterdijk's trilogy on "bubbles" is still being translated. This quote is from an exegesis of the German original by René ten Bos: "Towards an Amphibious Anthropology," 74–75. Bret Buchanan has traced the lateral movement of philosophical conversations about the umwelt since the time of Heidegger. Exploding lines of flight, ignored by Sloterdijk in his own citations, offer an alternative genealogy to the patrilineal heritage he claims as the "second coming" of Heidegger. Merleau-Ponty understood the umwelt in terms of a relational dynamic involving the coproduction of both animal and milieu. Relations of interanimality, within and between animals and their environments, make up bubbles of perception and action in the world of Merleau-Ponty. Deleuze and Guattari burst open these bubbles in their writings about organisms as assemblages—where the hour, the weather, hunger all go into producing affects in bodies. Buchanan, *Onto-Ethologies*, 134–35, 185.

20 Schatz, Beugnon, and Lachaud, "Time-Place Learning by an Invertebrate"; Schatz et al., "Prey Density and Polyethism."

21 Hölldobler and Wilson, *The Ants*, 367.

22 Hayward, "FingeryEyes."

23 Departing from the work of Emmanuel Lévinas, who pulled ethics away from abstractions and located ethical call and response within the living reality of the material world, Deborah Bird Rose accounts for interspecies responsibilities that are up close, face to face, in both life and death. If Lévinas ultimately rejected obligations to animals, creatures he regarded as lacking a "face," Rose develops her idea of ecological existentialism to think about how responsibility and accountability work across the species interface. Rose, *Wild Dog Dreaming*, 2–3.

24 Myers, "Performing the Protein Fold."

25 Roberto Keller, "Homology Weekly: Sensillae Trichoidea," Archetype blog, January 8, 2009, http://blog-rkp.kellerperez.com/2009/01/homology -weekly-sensillae-trichoidea /#more-179.

26 Reznikova, *Animal Intelligence*, 79–80.

27 Martin Giurfa, "Cognition with Few Neurons: Higher-Order Learning in Social Insects," Plenary Speech, International Congress, International Union for the Study of Social Insects, July 17, 2014, Cairns, Australia.

28 Schatz et al., "Prey Density and Polyethism."

29 Perhaps all organisms grasp the world with provisional maps (subject to revision) of shifting entanglements. Perhaps all conscious subjects build epistemological architectures, edifices of what Bruno Latour calls factishes, to mask the unknowable beyond. Factishes "create and are created by our practices," writes Latour's colleague and critic Isabelle Stengers. The limiting ring of phenomenological worlds, designated by the German preposi-

tion *um-* in umwelt, might be understood as an unstable architecture of factishes—a provisional map, subject to revision. Stengers, *Cosmopolitics I*, 36.

30 "Enemy of the Boll Weevil," *Houston Post*, May 28, 1904, Microfilm Publications M864, Records of the Division and Bureau of Entomology, 1863–1934, National Archives II, College Park, Maryland.

31 Cook, "An Enemy of the Cotton Boll Weevil," 863.

32 Haraway, "Speculative Fabulations for Technoculture's Generations"; Elden and Mendieta, "Being with as Making-Worlds," 6; Tsing, "Worlding the Matsutake Diaspora."

33 See Callon, "Some Elements of a Sociology of Translation"; Latour, *Science in Action*; Star and Griesemer, "Institutional Ecology."

34 "$250,000 for Cotton Investigations," July 20, 1904, and "Memorandum Concerning Allotment of Boll Weevil Appropriation, Fiscal Year Beginning July 1, 1905," Microfilm Publications M864, Records of the Division and Bureau of Entomology, 1863–1934, National Archives II, College Park, Maryland.

35 W. H. Hunter to L. O. Howard, July 14, 1904, Microfilm Publications M864, Records of the Division and Bureau of Entomology, 1863–1934, National Archives II, College Park, Maryland.

36 "Enemy of the Boll Weevil."

37 For cotton growers, "squares" are buds on the cotton plant that develop into flowers. W. H. Hunter to L. O. Howard, July 20, 1904, Microfilm Publications M864, Records of the Division and Bureau of Entomology, 1863–1934, National Archives II, College Park, Maryland.

38 Compare Foucault, *The History of Sexuality*, 139.

39 Cook, "The Social Organization and Breeding Habits," 46.

40 Even if initial ideas of interessement had an anthropocentric and entrepreneurial bias (as Collins and Yearley charged), they still have the potential to do new work in multispecies worlds. Collins and Yearley, "Epistemological Chicken," 313.

41 Hölldobler and Wilson, *The Ants*, 348; Cassil et al., "Cooperation during Prey Digestion."

42 Biologically active molecules are passed among ants in liquid food. Cues for nest mate recognition, as well as immune response, have been extensively studied. Adria LeBoeuf's talk given to the 2014 meetings of the International Union for the Study of Social Insects in Cairns, Australia, described her study of the transfer of juvenile hormone from adults to larvae in *Camponotus floridanus*. She found that feeding elevated levels of this hormone to larvae led them to develop larger heads. Compare other accounts of material-semiotic elements in Barad, *Meeting the Universe Halfway*; Haraway, *When Species Meet*; Hayward, "FingeryEyes."

43 Becker, "Doing Things Together"; Strauss, "A Social World Perspective," 119; Clarke, *Disciplining Reproduction*, 15.

44 Cook, "The Social Organization and Breeding Habits," 14.

45 Fujimura, "Authorizing Knowledge in Science and Anthropology," 347.

46 Here I am building on insights about scientific apparatuses by Shapin and Schaffer. I embedded certain assumptions in this apparatus—namely, that these ants would come to regard my assemblages of plastic and plaster as a nest and that their behavior in such a nest, exposed to the light of day, would be analogous to what they do underground. Shapin and Schaffer, *Leviathan and the Air-Pump*, 14, 112.

47 See Breed et al., "Thief Workers and Variation in Nestmate Recognition," 327; see also Breed et al., "Thievery, Home Ranges, and Nestmate Recognition." For a discussion of Breed's work, see chapter 1, note 26.

48 *Six Legs Better*, a cultural history of myrmecology (the scientific study of ants) by Charlotte Sleigh, offers a nuanced account of Wheeler's intellectual formation and his later battles with E. O. Wilson. See Sleigh, *Six Legs Better*, 248n4.

49 Sleigh, *Six Legs Better*, 79, emphasis added.

50 Sleigh, *Six Legs Better*, 169.

51 Millikan, "Pareto's Sociology," 327.

52 Quoted in Sleigh, *Six Legs Better*, 86.

53 Bataille, *The Accursed Share*, 21. Brenkman, "Introduction to Bataille," 61.

54 Stephen Pratt, who studied communication behavior in *Ectatomma* in the 1980s, described the sharing of liquid food in this species with loving attention to detail:

> Droplet-laden foragers returned immediately to the nest tube and, after a few seconds of excitation behavior, either stood still or walked slowly about the nest with [their] mandibles open and mouthparts usually retracted. They were generally approached within a few seconds by unladen workers who gently antennated the clypeus, mandibles, and labium of the drop-carrier, using the tips of their antennae. The carrier then opened its mandibles wide and pulled back its antennae, while the solicitor opened its mandibles, extruded its mouthparts and began to drink. During feeding, the solicitor continued to antennate the donor, who remained motionless. Usually the solicitor also rested one or both front legs on the head or the mandibles of the donor.

> Pratt, "Recruitment and Other Communication Behavior," 327.

55 See Breed et al., "Thievery, Home Ranges, and Nestmate Recognition."

56 Schuck-Paim, Pompilio, and Kacelnik, "State-Dependent Decisions Cause Apparent Violations."

57 Brenkman, "Introduction to Bataille," 62.

58 DeVries, "Enhancement of Symbioses between Butterfly Caterpillars and Ants"; DeVries and Baker, "Butterfly Exploitation of an Ant-Plant Mutualism."

59 Compare with Deleuze and Guattari, *A Thousand Plateaus*, 241–42; Kirksey and Helmreich, "The Emergence of Multispecies Ethnography," 546; Haraway, *When Species Meet*, 244.

60 Cook, "An Enemy of the Cotton Boll Weevil," 864.

61 Cook, "Professor William Morton Wheeler on the Kelep," 611.

62 Weber, "Two Common Ponerine Ants of Possible Economic Significance."

63 "Memorandum Concerning Allotment of Boll Weevil Appropriation, Fiscal Year Beginning July 1, 1905."

64 De la Fuente and Marquis, "The Role of Ant-Tended Extrafloral Nectaries"; Blüthgen et al., "How Plants Shape the Ant Community in the Amazonian Rainforest Canopy."

65 Marder, *Plant Thinking*, 85.

66 Perfecto and Vandermeer, "Cleptobiosis in the Ant *Ectatomma ruidum*."

67 Altshuler, "Novel Interactions of Non-pollinating Ants."

68 DeVries, "Enhancement of Symbioses between Butterfly Caterpillars and Ants"; DeVries and Baker, "Butterfly Exploitation of an Ant-Plant Mutualism."

69 Gruen, "Entangled Empathy."

70 Kockelman, "A Mayan Ontology of Poultry," 438–39.

71 Rose, *Reports from a Wild Country*, 27.

72 Kohn, *How Forests Think*, 75.

73 "The plant's absolute silence puts it in the position of the subaltern," writes Michael Marder. "Our incapacity to communicate with plants the way we do with other human beings (and even with certain animals) runs the risk of objectifying them or, at best, speaking *for* them, in their defense, if not in their place." In contrast to Jacob von Uexküll, who invited readers to "walk into unknown worlds" of animals, Marder insists we regard plants as entities that entail "profound obscurity," that flourish on the edge of visibility, known phenomena, and "the world." Marder asks, "How does the world appear (or not appear) to a plant? What is its relation to its world? What does it strive to, direct itself toward, or intend?" Marder, *Plant Thinking*, 9, 186.

74 Lévi-Strauss, *Totemism*, 63; Haraway, *When Species Meet*, 278.

75 Haraway, "Manifesto for Cyborgs," 118; Kohn, *How Forests Think*, 75.

THREE. HOPE IN THE REVERTED ZONE

1 Jon Fuller, "Re: Frog Pictures," e-mail sent to frognet@lists.frognet.org, accessed June 7, 2004, http://pets.dir.groups.yahoo.com/group/frognet/message/40816.

2 Drewry, Heyer, and Rand, "A Functional Analysis of the Complex Call," 636.

3 Compare Tsing, "Blasted Landscapes"; Kirksey, Shapiro, and Brodine, "Hope in Blasted Landscapes."

4 Lindsay-Poland, *Emperors in the Jungle*, 61–63; Lindsay-Poland, "U.S. Mili-

tary Base Contamination and Clean-Up in Panama," Fellowship of Reconciliation, February 1997, http://forusa.org/programs/panama/archives/0497 -1.htm.

5 Department of Defense, "Evaluation of Unexploded Ordnance Detection and Interrogation Technologies," Final Report: February 1997; prepared for Panama Canal Treaty Implementation Plan Agency, http://64.78.11.86 /uxofiles/enclosures/Panama_techeval.html#2.0.

6 Nature Conservancy and ANCON, "Rapid Ecological Assessment of the Lands in Panama Managed by the U.S. Department of Defense," unpublished report, submitted to Department of Defense on December 31, 1994.

7 Haraway, Modest_Witness, 104.

8 "Isthmian Explosive Disposal," accessed September 16, 2014, http:// uxopanama.wix.com/isthmian#!proyectos/ctwk.

9 The 1997 Department of Defense report that used environmental rhetoric to avoid paying for the cleanup of these sites invoked the presence of fourteen endangered animal species, as well as seven "critically imperiled" plant species, on Department of Defense lands in Panama. The Nature Conservancy followed designations under the U.S. Endangered Species Act. But six of these so-called endangered animals are now formally listed as being of "Least Concern" on the International Union for Conservation of Nature's (IUCN) Red List of Endangered Species: the brown pelican (*Pelecanus occidentalis*), the wood stork (*Mycteria americana*), the peregrine falcon (*Falco peregrinus*), the howler monkey (*Alouatta palliata*), the ocelot (*Leopardus pardalis*), and the jaguarundi (*Herpailurus yagouaroundi*). Two animals living on these lands are near threatened: the margay (*Leopardus wiedii*) and the jaguar (*Panthera onca*). The Department of Defense report actually found only two endangered land-dwelling animals—the spider monkey (*Ateles geoffroyi*) and Baird's tapir (*Tapirus bairdii*)—and one critically endangered animal: the cotton-top tamarin (*Saguinus oedipus*). Aquatic animals, which presumably would not be seriously disturbed by clearing unexploded ordnance, were also listed in the report: the near-threatened South American river otter (*Lontra longicaudis*) as well as the vulnerable West Indian manatee (*Trichechus manatus*) and the American crocodile (*Crocodylus acutus*). See Department of Defense, "Evaluation of Unexploded Ordnance Detection and Interrogation Technologies"; Nature Conservancy and ANCON, "Rapid Ecological Assessment of the Lands in Panama Managed by the U.S. Department of Defense."

10 Compare Masco, "Mutant Ecologies."

11 Derrida, "Plato's Pharmacy," 97, 99; Stengers, *Cosmopolitics I*, 29.

12 Geoghegan, "An Anti-humanist Utopia?," 50.

13 Geoghegan, "An Anti-humanist Utopia?," 49–51.

14 Ronan, "Observations on the Word Gringo."

15 Carse, "Nature as Infrastructure"; Ahmed, *The Promise of Happiness*, 44.

16 Reviewing Derrida's work, acclaimed literary critic Fredric Jameson writes,

"We ought to be able to distinguish an apocalyptic politics from a messianic one, and which might lead us on into some new way of sorting out the Left from the Right, the new International in Marx's spirit from that in the world of business and state power." Jameson, "Marx's Purloined Letter," 63–64.

17 "The very idea of the messianic brings the whole feeling of dashed hopes and impossibility along with it," according to Jameson. "You would not evoke the messianic in a genuinely revolutionary period, a period in which changes can be sensed at work all around you." Jameson, "Marx's Purloined Letter," 62.

18 Derrida, "Marx and Sons," 253.

19 Derrida, *Specters of Marx*, 74.

20 Benjamin, *Illuminations*, 262.

21 Palaversich, *De Macondo a McOndo*.

22 Harvey, *Spaces of Hope*, 177, 195.

23 Kirksey, Shapiro, and Brodine, "Hope in Blasted Landscapes," 39.

24 Michel Foucault would have had a field day if he had been given the opportunity to visit the City of Knowledge. Hegemonic visions of medicalization, governmentality, and biopolitics were being implemented by these transnational organizations in a place where knowledge was intimately entangled with power. See, for example, Foucault, *Power/Knowledge*; Foucault, *The Birth of the Clinic*.

25 Hartigan, "Millenials [sic] for Obama"; Rutherford, *Laughing at Leviathan*.

26 Berlant, *Cruel Optimism*, 24.

27 Berlant, *Cruel Optimism*, 48.

28 Harvey, *Spaces of Hope*, 23; Stengers, "The Cosmopolitical Proposal."

29 Thompson, "The Privatization of Hope and the Crisis of Negation," 9.

30 "About 30% (1,895) of the 6,285 amphibian species assessed by the IUCN are threatened with extinction," according to the Amphibian Ark website. "There are 6% (382) known to be Near Threatened and 25% (1,597) are data deficient. This means that about 3,900 species are in trouble." "Frightening Statistics," Amphibian Ark, accessed September 16, 2014, http://www .amphibianark.org.

31 Haraway, *When Species Meet*, 4.

32 "Figures collect the people through their invitation to inhabit the corporeal story told in their lineaments," writes Haraway. "Figures are not representations or didactic illustrations," she argues, "but rather material-semiotic nodes or knots in which diverse bodies and meanings coshape one another." Like metaphors, figures are not literal and self-identical— they diffract our visions. Figures exist in the borderlands between metaphorical allegory and sacramental actuality. Substances and symbols come together in figures to coshape and coconstitute institutions, bodies, and worlds. By pinning their hopes on specific figures, people can bring substantive objects within the reach of desire. Haraway, *When Species*

Meet, 4; see also Bateson, *Steps to an Ecology of Mind*, 33–37; Haraway, *Modest_Witness*, 11.

33 Leibniz quoted in Ahmed, *The Promise of Happiness*, 92.

34 Berlant, *Cruel Optimism*, 48.

35 The genus *Atelopus* has 113 described species. Approximately sixty to seventy-five of all known *Atelopus* species have experienced rapid population declines. This facility in the Reverted Zone was focused on breeding four species of *Atelopus*: *A. limosus*, *A. glyphus*, *A. certus*, and *A. zeteki*, the Panamanian golden frog. The collection sites were Cerro Pirre, Cerro Bruja, and Cerro Sapo in Darien as well as Rio Mamoni, a popular whitewater rafting destination. For updated information, see the Smithsonian's Amphibian Rescue and Conservation Project website at http://amphibianrescue.org.

36 Sandrine Ceurstemont, "Endangered Tree Frog Bred for the First Time," New Scientist TV, November 18, 2010, http://www.newscientist.com/blogs/nstv/2010/11/endangered-tree-frog-bred-for-the-first-time.html.

37 Rose and van Dooren, "Unloved Others."

38 Haraway, "Speculative Fabulations for Technoculture's Generations."

39 Mary Catherine Bateson has written a brilliant essay titled "Turning into a Toad" on the limits of our ability to empathize with amphibians and other endangered species. Bateson, *Peripheral Visions*, 127–43.

40 Kohler, *Lords of the Fly*, 45.

41 Sara Ahmed has described other situations where people feel alienated from dominant affects. "We can feel alienated by forms of happiness that we think are inappropriate," Ahmed writes. "Take the example of laughter in the cinema. How many times have I sunk desperately into my chair when that laughter has been expressed at points I find far from amusing" (*The Promise of Happiness*, 42). In the moment when the U.S. troops left Panama, the Wounaan were similarly alienated from the dominant affect of postcolonial national liberation.

FOUR. HAPPINESS AND GLASS

1 Lucy Cook, "2010: A Frog Odyssey," Amphibian Avenger, accessed September 16, 2014, http://pinktreefrog.typepad.com.

2 For comparison, see Rose and van Dooren, "Unloved Others."

3 Hopes emerge with practices of "care for the future," writes Thom van Dooren. Writing of an ark to keep endangered snails alive in Hawaii, van Dooren has described the regimes of dedicated daily care and the ongoing curiosity about how to make the conditions, and indeed the lives of snails, better. Thick notions of care, according to Maria Puig de la Bellacasa, require that it be understood as simultaneously "a vital affective state, an ethical obligation and a practical labor." Van Dooren, "Extinction at a Snail's Pace," Hawaii Forest and Trail, October 10, 2010, http://www.hawaii

-forest.com/index.php/articles/extinction-at-a-snails-pace; Puig de la Bellacasa, "Nothing Comes without Its World," 197; Haraway, "A Manifesto for
Cyborgs."

4 F. Solís, R. Ibáñez, J. Savage, C. Jaramillo, Q. Fuenmayor, B. Kubicki, J.
Pounds, G. Chaves, K. Jungfer, and K. Lips, *Agalychnis lemur*," IUCN Red
List of Threatened Species, Version 2014.3, 2008, http://www.iucnredlist
.org.

5 "Scientists Leap to Save Golden Frog in Panama," *Washington Post*, November 7, 2006.

6 Stuart et al., "Status and Trends of Amphibian Declines and Extinctions
Worldwide."

7 Berlant, *Cruel Optimism*, 48.

8 Ahmed, *The Promise of Happiness*, 43–46, 92; Berlant, *Cruel Optimism*, 48.

9 For a discussion of oblique powers in Latin America, see García Canclini,
Hybrid Cultures.

10 "AArk Activities," Amphibian Ark, accessed April 27, 2015, http://www
.amphibianark.org.

11 An edited video of this speech by Kevin Zipple has been posted on YouTube.
Amphibian Ark, "Our Planet's Canaries in the Coal Mines," YouTube, July
25, 2011, http://www.youtube.com.

12 Harding, "Get Religion," 354.

13 Derrida's hopeful sense of expectation is not oriented toward a specific
Messiah. Instead, Derrida celebrates a quasi-transcendental force he calls
"messianicity without messianism." In contrast to Christian messianic
traditions, which pin hopes to a particular figure (Jesus Christ), Derrida's
notion of messianicity is "without content." Celebrating messianic desires
that operate beyond the confines of any particular figure, his notion of the
messianic spirit involves a universal structure of feeling that works independently of any specific historical moment or cultural location. Derrida,
"For a Justice to Come"; Derrida, *Specters of Marx*, 167; Derrida, "Marx and
Sons," 253.

14 David Wood suggests that "Derrida is articulating a practice (and a patience) of creative, critical openness predicated on the recognition, simple
but profound, that the future never comes 'as such.' What comes is always
another moment with its own mixture of horizons, uncertainties, exposure, and impossibility, its own manner of taking up/taking on the past and
of opening onto 'the future.' This open, differential, eventuating matrix of
thinking and engagement is not a substitute for concrete political strategy
either, but it functions as a source of recursive resistance to both dogma-
tism and despair" ("On Being Haunted by the Future," 280). However, I
contend that careful attention to Derrida's original texts reveals that he
was sharply critical of "concrete political strategy" in nationalist movements, for example. The concreteness of Zipple's interventions would likely
trouble Derrida. Avoiding these sorts of attachments, the spirit of Derrida's

thought instead dances alone in an imaginative desert. Hoping for nothing in particular, harboring the empty promise of Derrida's messianicity, has no future—it is literally pointless and goes nowhere. For a more comprehensive engagement with Derrida's writing on the messianic, see Kirksey, *Freedom in Entangled Worlds*, 25. For an application of these ideas to the BP Deepwater Horizon explosion, see Kirksey, Shapiro, and Brodine, "Hope in Blasted Landscapes," 236.

15 Messianic hopes in the biosciences are often problematic since they often involve what Donna Haraway describes as "misplaced concreteness." Biotech ventures have been critiqued for using messianic discourse to focus hopes of researchers, venture capitalists, and consumers on things that are too specific—like a gene, or a new pharmaceutical drug, or the resurrected cells of an extinct species. Kevin Zipple is revoicing speculative fictions and fabulations that have emerged at the intersection of biological sciences and economic enterprises—turning the rhetorical power of messianic discourse from producing profit for humans to producing future generations of organisms he loves. Haraway, *Modest_Witness*, 147. See also Cheah, "Spectral Nationality," 248; Haraway, "Speculative Fabulations for Technoculture's Generations"; Fortun, "Mediated Speculations in the Genomics Futures Markets," 141; Sunder Rajan, *Biocapital*, 123.

16 Extinctions were taking place along a dull edge of time, in the words of Thom van Dooren, with a drawn-out and ongoing process of loss taking place long before and well after the final death. Van Dooren, *Flight Ways*, chapter 2.

17 Benjamin, *Illuminations*, 264.

18 Ahmed, *The Promise of Happiness*, 43–46, 92; Berlant, *Cruel Optimism*, 48.

19 Sloterdijk, *Bubbles*, 48.

20 Ahmed, *The Promise of Happiness*, 42.

21 Chrulew, "Managing Love and Death at the Zoo," 137–39.

22 See van Dooren, *Flight Ways*, chapter 4.

23 Vicky Poole, "Husbandry Manual: Panamanian Golden Frog," Project Golden Frog, 2006, 3–4, http://www.ranadorada.org/PDF/Husbandry Manual.pdf.

24 The Panamanian golden frogs were technically placed "on loan" from the Maryland Zoo. Only institutions in the Association of Zoos and Aquariums were eligible to participate in the lending program, according to guidelines established by the U.S. Fish and Wildlife Service. "This restriction was intended to prevent the protected species from entering the pet trade via captive zoo breeding," writes Vicky Poole, who authored the "Husbandry Manual" for the Panamanian golden frog while working for the National Aquarium in Baltimore. "Wild-caught illegal specimens could be 'laundered,'" according to Poole, "under the guise of coming from legal 'zoo stock'" ("Husbandry Manual," 3). I conducted a series of interviews with Poole and consulted her husbandry manual.

25 "Long-term captive management plan is to maintain 30–50 frogs from each bloodline," according to the *Atelopus zeteki* "Husbandry Manual." Frogs from some mating pairs "will be undesirables and may displace other more valuable offspring from a desirable breeding, so euthanasia will be necessary to eliminate, or at least reduce their numbers. Be prepared to house offspring indefinitely if allowed to survive." Poole, "Husbandry Manual," 13.

26 Van Dooren, *Flight Ways*, 117.

27 Foucault, "Of Other Spaces," 26, 25. See also Khan, "Heterotopic Dissonance in the Museum Representation of Pacific Cultures"; Bennett, "The Exhibitionary Complex."

28 Van Dooren, *Flight Ways*, 117.

29 Williams, "Utopia and Science Fiction," 212.

30 Stengers, *Cosmopolitics II*, 347.

31 See Lévi-Strauss, *The Savage Mind*, 21; de Certeau, *The Practice of Everyday Life*; da Costa and Philip, *Tactical Biopolitics*, xvii.

32 Ordinary citizens who keep frogs and reptiles as pets now have ready access to the same genetically modified animals that have long been used by zoos and specialized breeding facilities. These adaptable insects are proliferating in biotechnical worlds and emergent ecologies orbiting around the pet industry. Kohler, *Lords of the Fly*, 45.

33 Kowal and Radin, "Indigenous Biospecimen Collections and the Cryopolitics of Frozen Life."

34 Da Costa and Philip, *Tactical Biopolitics*; Kirksey, Schuetze, and Helmreich, "Introduction," 9.

35 Rather than Jeremy Bentham's Panopticon, where someone might or might not be observing from a figural guard tower, we worked to re-create Bruno Latour's Oligopticon—where a small group of people can accurately monitor an object of interest. Foucault, *Discipline and Punish*, 195; Latour, *Reassembling the Social*, 181.

36 *The Multispecies Salon* involved collaborations among ethnographers and artists who explored a series of interrelated questions: "Which beings flourish, and which fail, when natural and cultural worlds intermingle and collide? What happens when the bodies of organisms, and even entire ecosystems, are enlisted in the schemes of biotechnology and the dreams of biocapitalism? And finally, in the aftermath of disasters—in blasted landscapes that have been transformed by multiple catastrophes—what are the possibilities of biocultural hope?" Kirksey, Schuetze, and Helmreich, "Introduction," 1.

37 "Studbooks," Association of Zoos and Aquariums, accessed February 17, 2014, http://www.aza.org/studbooks.

38 Stengers, *Cosmopolitics II*, 347.

39 See Marcus, *Para-sites*, 5.

40 "Frightening Statistics," Amphibian Ark, accessed March 3, 2014, http://www.amphibianark.org.

41 Berlant, *Cruel Optimism*, 48.

42 Kirksey, Shapiro, and Brodine, "Hope in Blasted Landscapes," 39.

43 Ahmed, *The Promise of Happiness*, 31, 186.

44 Ahmed, *The Promise of Happiness*, 22.

45 The work of both Sarah Franklin and Sara Ahmed has been critical in help-ing me make this point.

46 Taussig, *Shamanism, Colonialism, and the Wild Man*, 212.

47 Lawrence, *Rodeo*, 7.

48 Franklin, "Ethical Biocapital," 102.

49 "Wild," Urban Dictionary, accessed September 18, 2014, http://www
.urbandictionary.com.

50 For a discussion of mutant fruit flies running wild in the realms of biosci-ence and bioart, see Kirksey, Costelloe-Kuehn, and Sagan, "Life in the Age
of Biotechnology," 197.

51 Collard, "Putting Animals Back Together, Taking Commodities Apart," 154.

52 Anonymous interview, Gamboa, Panama, February 17, 2014.

53 Stengers, *Cosmopolitics II*, 35–36.

54 See Shukin, *Animal Capital*, 225–26.

55 Quotes taken from telephone interview with Vicky Poole and Kevin Barrett,
September 5, 2014.

56 Haraway, "Speculative Fabulations for Technoculture's Generations."

57 "Chytridiomycosis," Amphibian Ark, accessed March 3, 2014, http://www
.amphibianark.org.

58 Bletz et al., "Mitigating Amphibian Chytridiomycosis with Bioaugmenta-tion," 813; Baitchman and Pessier, "Pathogenesis, Diagnosis, and Treatment
of Amphibian Chytridiomycosis," 680.

59 In February 2014, I interviewed Myra Hughey at the Smithsonian Tropical
Research Institute, who was collaborating with Reid Harris at James Madi-son University. The work by Hughey and Harris was still in its early stages
at this time.

60 "Biotechnology occupies a messianic space," writes Kaushik Sunder Rajan,
"of technology and of Life linked through capital." Hopes by human pa-tients, who are waiting for new cures, often become mixed up with specula-tion by entrepreneurs who dream about cashing in on new miracle treat-ments. Money-making schemes can nonetheless bring interesting things to
life. Sunder Rajan, *Biocapital*, 149, 123.

61 Fortun, "Mediated Speculations," 146.

62 For an account of ethics, messianic thought, and pharmaceutical drug de-velopment, see Kirksey, Costelloe-Kuehn, and Sagan, "Life in the Age of
Biotechnology," 188–89.

63 For an account of symbolic capital, see Bourdieu, *Distinction*.

FIVE. BUBBLES

1 Sloterdijk, *Bubbles*, 12.

2 Sloterdijk, *Bubbles*, 11.

3 For a critical discussion of swarming in the context of recent work by Hardt and Negri, Jake Kosek, Hugh Raffles, and Eugene Thacker, see Kirksey et al., "Poaching at the Multispecies Salon."

4 This point is explored with respect to chytrids, frogs, fig trees, and wasps in Kirksey, "Species."

5 According to a 2006 article in *Nature* by Timothy James, chytrids resemble the ancestors of fungi. Genetic evidence suggests that all fungi once had aquatic spores that swam around with whip-like flagella. But not everyone agrees. When I wrote a casual e-mail to Dorion Sagan, a friend and collaborator, saying that I was about to "do some deep hanging out with one of the few experts on chytrid fungi," he wrote right back, saying, "They are not really fungi." Pointing me to a paper by his mother, Lynn Margulis, he claimed they are Protoctista—part of "a messy kingdom, containing by definition what is not fungi, plants, animals, or monera (bacteria and archaea)." Still, everyone agrees on some basic facts: these single-celled organisms, which look much like human sperm, have a cell nucleus and many of the basic organelles of multicellular organisms, eukaryotes. James et al., "Reconstructing the Early Evolution of Fungi."

6 Ten Bos, "Towards an Amphibious Anthropology," 74.

7 Von Uexküll, *A Foray into the Worlds of Animals and Humans*, 42; Agamben, *The Open*, 40–41. See chapter 2, note 12.

8 Ten Bos, "Towards an Amphibious Anthropology," 75.

9 Stengers, *Cosmopolitics II*. On co-option, see also Ingold, *The Perception of the Environment*, 175.

10 Chytrid cells have a nucleus, like multicellular organisms, as well as mitochondria (the powerhouse organelles of animals and plants). Some chytrids live in oxygen-free environments and have hydrogenosomes, organelles that produce hydrogen.

11 Sagan is circumspect when he describes the phenomenological worlds of bacteria. Expanding beyond conventional boundaries separating living beings from physical matter to discuss poltergeists and thermodynamic processes in an essay about umwelten, he concludes on an ambivalent note: "It is too early to say definitively who, or what, does and does not have an Umwelt." Sagan, "Umwelt after Uexküll," 27; Moss et al., "Chemotaxis of the Amphibian Pathogen *Batrachochytrium dendrobatidis*."

12 Compare with Deleuze and Guattari, *A Thousand Plateaus*, 161.

13 For Karen Barad's notion of intra-action, see Barad, "Meeting the Universe Halfway," 815; Barad, "Invertebrate Visions," 238n20.

14 Barad, "Meeting the Universe Halfway," 817.

15 Voyles et al., "Pathogenesis of Chytridiomycosis," 582–85.

16 Swei et al., "Is Chytridiomycosis an Emerging Infectious Disease in Asia?"; Jodi Rowley, personal communication, November 15, 2012.

17 Rosenblum et al., "Complex History of the Amphibian-Killing Chytrid Fungus Revealed."

18 For more on torque, see Bowker and Star, *Sorting Things Out*.

19 For more on this point, see Kirksey, "Species."

20 Simmons and Longcore, "*Thoreauomyces* gen. nov., *Fimicolochytrium* gen. nov. and Additional Species in Geranomyces."

21 Schloegel et al., "Novel, Panzootic and Hybrid Genotypes of Amphibian Chytridiomycosis."

22 MacNeill, "Parasexuality in Plant Pathogenic Fungi," 147.

23 Monstrous mutations in animals and plants usually result in the death of the organism, but "hope" emerges when strange new forms occasionally survive and give rise to new species. Stephen Jay Gould suggests that "new species arise abruptly by discontinuous variation, or macromutation." Gould, "Return of Hopeful Monsters," 24.

24 Rosenblum, "Complex History of the Amphibian-Killing Chytrid Fungus Revealed," 1.

25 Rotman, *Becoming Beside Ourselves*, 104.

26 Serres, *The Parasite*, 161–62.

27 Sloterdijk, *Bubbles*, 48.

28 Sloterdijk, *Bubbles*, 51.

29 See Balakrishnan and Schmitt, *The Enemy*, 108; Kirksey, *Freedom in Entangled Worlds*, 177.

30 Sloterdijk, *Bubbles*, 46.

31 Donna Haraway has played with the language of labor to write about situations where "workers" are not all human persons with contractual rights, but also nonhuman animals. She insists we take "animals seriously as workers without the comforts of humanist frameworks for people or animals." Haraway, *When Species Meet*, 73.

32 Sloterdijk, *Bubbles*, 23.

33 Charis Thompson (née Cussins) coined the phrase "ontological choreography" to describe the coordinated action of many actors. I use the word "choreography" here, following Thompson, "to invoke materiality, structural constraint, performativity, discipline, co-dependence of setting and performers, and movement." Cussins, "Ontological Choreography," 604; Thompson, *Making Parents*.

34 Karen Barad's notion of intra-action (in contrast to the usual "interaction," which presumes the prior existence of independent entities/relata) represents a profound conceptual shift. It is through specific agential intra-actions that the boundaries and properties of the "components" of phenomena become determinate. "Intra-actions are constraining but not determining. . . . 'We' are not outside observers of the world. Nor are we simply located at particular places in the world; rather, we are part of the

world in its ongoing intra-activity. . . . What possibilities exist for agency, for intervening in the world's becoming? Where do the issues of responsibility and accountability enter in?" Barad, "Meeting the Universe Halfway," 815, 824, 826, 828.

SIX. XENOECOLOGIES

1 Gurdon, "*Xenopus* as a Laboratory Animal"; Weldon et al., "Origin of the Amphibian Chytrid Fungus"; Tinsley, "Parasites of *Xenopus*."

2 Tinsley, Chappell, and Kobel, "Geographical Distribution and Ecology," 44–50.

3 Tinsley, Chappell, and Kobel, "Geographical Distribution and Ecology," 44–50.

4 Elkan, "The Xenopus Pregnancy Test," 1253.

5 Elkan, "The Xenopus Pregnancy Test," 1255. Controversies about who "discovered" the *Xenopus* pregnancy test were mired in colonial politics. In Great Britain it became known as the Hogben Test. It was named after Lancelot Hogben—a British zoologist and medical statistician who was a prominent critic of eugenics and who founded a new field of "social biology" at the London School of Economics. In 1934, Hogben published a note in the journal *Nature*, where he reported the discovery of the test. A South African researcher, namely J. W. C. Gunn, wrote a letter to the *British Medical Journal* contesting Hogben's claim to the discovery. According to Gunn, one of his colleagues at the University of Capetown, a Dr. Zwarenstein, deserves to be credited for the test. See Hogben's response, "*Xenopus* Test for Pregnancy," 38.

6 Gordon, *Woman's Body, Woman's Right*, 204.

7 "'DIY Frog Pregnancy Test' at Proteus Gowanus: Women to Inject Urine into Frogs to See If They're Pregnant," *Huffington Post*, June 26, 2012, http://www.huffingtonpost.com.

8 Leavitt, "A Private Little Revolution," 322; see also Olszynko-Gryn, "When Pregnancy Tests Were Toads," 2.

9 Elkan to Sanger, September 17, 1951, Series II (Subseries 1, correspondence), MSP#: 211602, Sophia Smith Collection, Margaret Sanger Papers, Microfilm Collection, Bobst Library, NYU.

10 Herrel and van der Meijden, "An Analysis of the Live Reptile and Amphibian Trade," 104.

11 Weldon et al., "Origin of the Amphibian Chytrid Fungus"; Vredenburg et al., "Prevalence of *Batrachochytrium dendrobatidis* in *Xenopus*."

12 See Wald, *Contagious*.

13 The familiar phrase "There is always something new coming out of Africa" originated in Greece, no later than the fourth century BC. Whereas "something new" meant strange hybrid animals to Aristotle, twentieth-century writers and filmmakers began using the phrase with a sense of admiration. Feinberg and Solodow, "Out of Africa," 255.

14 For an account of other artworks in this exhibit, and a discussion of the curatorial impulses animating the show, see Kirksey, *The Multispecies Salon*.

15 Kirksey et al., "The Xenopus Pregnancy Test," n.p.

16 The ad was posted on *Brokelyn*, http://brokelyn.com/.

17 "'DIY Frog Pregnancy Test.'"

18 Cassandra Garrison, "Brooklyn Gallery Wants Women to Inject Frogs with Urine to See If They're Pregnant," *Metro*, June 25, 2012, http://www.metro.us.

19 Sarah Fecht, "From Living Room to Lily Pad: Is the Fatal Amphibian Chytrid Fungus Spread via Pet Frogs?," *Scientific American*, July 20, 2012, http://www.scientificamerican.com.

20 "Cruel Exhibit Stopped after PETA Steps In," PETA Blog, July 2, 2012, http://www.peta.org.

21 Cassandra Garrison, "UPDATE: Brooklyn Gallery Cancels Frog Pregnancy Tests after PETA Outrage," *Metro*, June 27, 2012, http://www.metro.us.

22 "Products," Xenopus Express, accessed April 27, 2015, http://www.xenopus.com.

23 Haraway, *When Species Meet*, 70.

24 Green, *The Laboratory Xenopus sp.*, 110, 114.

25 Elepfandt, "*Xenopus* Sensory Perception," 114–16.

26 Compare Bennett, *Vibrant Matter*; Barad, "Invertebrate Visions."

27 Haraway, *Modest_Witness*, 79.

28 Haraway, *The Companion Species Manifesto*, 15.

29 Haraway, "Manifesto for Cyborgs," 65.

30 Wolfe, *What Is Posthumanism?*

31 Haraway, *When Species Meet*, 70; Kirksey et al., "The Xenopus Pregnancy Test."

32 Wolfe, *What Is Posthumanism?*, xxv.

33 See Kirksey et al., "The Xenopus Pregnancy Test." See also Mol, *The Body Multiple*; Thompson, *Making Parents*; Barad, "Meeting the Universe Halfway."

34 Cussins, "Ontological Choreography," 600.

35 Thompson, *Making Parents*, 8.

36 Gurdon, "*Xenopus* as a Laboratory Animal," 4.

37 NASA, "Spacelab-J/STS-47: Mission Overview," accessed April 3, 2014, http://lis.arc.nasa.gov.

38 Green, *The Laboratory Xenopus sp.*, xii.

39 "By the late 20th century, our time, a mythic time, we are all chimeras, theorized, and fabricated hybrids of machine and organism," according to Haraway's influential manifesto. "In short, we are cyborgs. This cyborg is our ontology; it gives us our politics." Haraway, "Manifesto for Cyborgs," 66.

40 Tinsley and McCoid, "Feral Populations of *Xenopus* outside Africa," 81.

41 Having extra sets of chromosomes, or polyploidy, is common in plants but

relatively rare in animals since it can disrupt the cellular dynamics that make sex possible. Most frogs are diploid, but a number of amphibians, like *Hyla versicolor* and *Ceratophrys ornata*, are polyploid. Encyclopaedia Britannica, *The Encyclopaedia Britannica Guide to Genetics*, 183; Duellman and Trueb, *Biology of Amphibians*, 451; Evans et al., "A Mitochondrial DNA Phylogeny of African Clawed Frogs."

42 In addition to finding *Bd* infections in 89.7 percent of *Xenopus* frogs in the Northern Cape of South Africa, Weldon also did limited swabbing in two populations living in the wilds of California, where he found infection rates of 3 percent and 14.3 percent. Weldon, "Chytridiomycosis," 2.

43 Garmyn et al., "Waterfowl," e35038.

44 McMahon et al., "Chytrid Fungus," 210.

45 Swei et al., "Is Chytridiomycosis an Emerging Infectious Disease in Asia?," e23179.

46 Rosenblum, "Complex History of the Amphibian-Killing Chytrid Fungus," 1.

47 Vredenburg et al., "Prevalence of *Batrachochytrium dendrobatidis* in *Xenopus*."

48 Davidson et al., "Anuran Population Declines Occur on an Elevational Gradient," 503.

49 Myra Hughey, "Diversity and Symbiosis: Examining the Taxonomic, Genetic, and Functional Diversity of Amphibian Skin Microbiota," poster on display at the Smithsonian Tropical Research Institute, Panama City, January 2014.

50 Serres, *The Parasite*, 7, 15–16, 253.

51 Comaroff and Comaroff, "Naturing the Nation," 251.

52 "Killer Meat-Eating Frogs Terrorize San Francisco," Fox News, March 14, 2007, http://www.foxnews.com.

53 Phillip Matier and Andrew Ross, "The Killer Frogs of Lily Pond," *San Francisco Chronicle*, March 12, 2007, http://www.sfgate.com.

54 The question of when to kill, and not kill, is tricky in multispecies worlds. Donna Haraway refuses the command from the Christian God, "Thou shalt not kill," while proposing a commandment of her own, "Thou shalt not make killable." Haraway, *When Species Meet*, 80.

55 Haraway, "The Promises of Monsters," 324.

56 Matier and Ross, "The Killer Frogs of Lily Pond."

57 Deborah Bird Rose has written of ethics of killing that channel death back into "the future complexity of life." Rose, "What If the Angel of History Were a Dog?," 77.

58 Stengers, *Cosmopolitics II*, 373.

59 "What Is a Nonnative Species?," Florida Fish and Wildlife Conservation Commission, accessed November 11, 2013, http://myfwc.com.

60 Layne, "Nonindigenous Mammals," 163.

1   Mooallem, "What's a Monkey to Do in Tampa?"
2   "Mystery Monkey of Tampa Bay," Facebook, accessed September 18, 2014, https://www.facebook.com/MysteryMonkeyOfTampaBay.
3   Lestel and Gaunet, "Etho-ethnology and Ethno-ethology"; Matsutake Worlds Research Group, "A New Form of Collaboration."
4   Cormier, *Kinship with Monkeys*; Fuentes, "Ethnoprimatology and the Anthropology of the Human-Primate Interface"; Riley, "Contemporary Primatology in Anthropology."
5   During our collective fieldwork in Florida, we shared video recordings, written field notes, observational data, and comments on our various manuscripts in progress. I spent three months on this project in January 2008, January 2012, and January–February 2013. Elan Abrell spent three weeks in the field in January–February 2013, while conducting research for his PhD at the CUNY Graduate Center. Tiffany Wade spent five months on the project in 2012 and 2013 collecting data for her master's thesis at San Diego State University. Erin Riley conducted two months of research in 2012 and 2013 for a coauthored manuscript with Tiffany Wade. We are still working on a future publication that will integrate our understandings of macaque feeding ecology, Florida's animal sanctuary movement, and the ethical affordances of boating technologies on the Silver River.
6   Deleuze and Guattari, *A Thousand Plateaus;* Haraway, *When Species Meet*; Kirksey and Helmreich, "The Emergence of Multispecies Ethnography."
7   Rose, *Reports from a Wild Country*, 4; compare Haraway, *Primate Visions*, chapter 3.
8   Hollander, *Raising Cane in the 'Glades*; Benitez Rojo and Maraniss, *The Repeating Island*.
9   Franklin, "Ethical Biocapital," 102.
10 See Franklin, "Cyborg Embryo."
11 Maestripieri, *Macachiavellian Intelligence*.
12 The phrase "inner frontier" originates with Robert Kohler. Ogden, *Swamplife*, 110–12; Kohler, *All Creatures*.
13 Parreñas, "Producing Affect."
14 See Rose, *Reports from a Wild Country*, 213; Haraway, "Speculative Fabulations for Technoculture's Generations," 253.
15 Candea, "We Both Wait Together."
16 My use of the notion of the "wild" departs from many scholars who have critically discussed this word over the past three decades. In particular, see Plumwood, *Feminism and the Mastery of Nature*; Cronon, *The Trouble with Wilderness*; Braun, *The Intemperate Rainforest*; Whatmore and Thorne, "Wild(er)ness."
17 Collard, "Putting Animals Back Together, Taking Commodities Apart," 154.
18 Constantino, *Tortoise Soup for the Soul*, 90.

19 Morel, "Trappers Use Fruit, Sandwiches."

20 Fjellman, *Vinyl Leaves*.

21 Wild Things, in a phrase, had institutionalized "the exhibitionary complex," the constellation of forces described by Tony Bennett as a vehicle "for inscribing and broadcasting the messages of power throughout society." The museums and exhibitions originally described by Bennett broadcast messages about the past, evolution, aesthetics, and man. Bennett, "The Exhibitionary Complex," 73.

22 Herrel and van der Meijden, "An Analysis of the Live Reptile and Amphibian Trade," 105.

23 Collard, "Putting Animals Back Together, Taking Commodities Apart."

24 Thanks are due to Elan Abrell for helping me describe and analyze the dynamics at Wild Things.

25 Elmore and Eberle, "Monkey B Virus"; Kessler and Hilliard, "Seroprevalence of B Virus (*Herpesvirus simiae*) Antibodies."

26 Engel et al., "Human Exposure to Herpesvirus B–Seropositive Macaques."

27 Whitney Rugg, "Kitsch," Chicago School of Media Theory, accessed September 18, 2014, http://csmt.uchicago.edu.

28 Ogden, *Swamplife*, 110–12.

29 Adams King, "Through the Looking Glass of Silver Springs," 2–3.

30 Vivanco and Gordon, *Tarzan was an Eco-tourist*.

31 Hamaker, "Springs Monkeys Preceded Tarzan."

32 Wolfe and Peters, "History of the Freeranging Rhesus Monkeys."

33 I first became aware of Paradise Park's history through a fortuitous meeting with Reginald Lucas, who grew up nearby and remembered visiting there as a child. Reginald now works as the mail facility supervisor at the CUNY Graduate Center, where he also provided critical logistical support during the research and writing of this book.

34 One hundred and forty-nine known lynchings of blacks were recorded in Florida during the first half of the twentieth century, from 1900 to 1945. Howard, "Vigilante Justice and National Reaction," 32–33.

35 Bradburd, "Adventure in the Zeitgeist, Adventures in Reality," 51; Haraway, *Primate Visions*, 403n18.

36 Compare Collard, "Putting Animals Back Together."

37 Haraway, *Primate Visions*, 41.

38 Gottschalk, "Study of the Silver River Monkeys."

39 Bernstein and Ehardt, "Intragroup Agonistic Behavior in Rhesus-Monkeys."

40 See, e.g., Thierry, Mewa, and Kaumanns, *Macaque Societies*; Silk, "Male Bonnet Macaques Use Information."

41 Datta, "The Acquisition of Dominance."

42 Datta, "The Acquisition of Dominance."

43 Compare Ogden, *Swamplife*, 112.

44 Cassidy and Mullin, *Where the Wild Things Are Now*, 278.

45 Knight, *Animals in Person*, 247.

46   Compare Parreñas, "Producing Affect."

47   Compare Sloterdijk, *Spheres*.

48   David Fleshler, "A Record Run: No Fatal Gator Attack in 5 Years," *Sun Sentinel*, October 28, 2013, http://articles.sun-sentinel.com.

49   Thanks are due to Tiffany Wade and Erin Riley for help in analyzing this encounter.

50   Massumi, *Parables for the Virtual*, 26; Parreñas, "Producing Affect."

51   Compare Moore and Kosut, *Buzz*.

52   Parreñas, "Producing Affect."

53   These field notes were produced with help from Tiffany Wade, who conducted a preliminary analysis of our collective video and audio recordings.

54   Compare de Certeau, *The Practice of Everyday Life*; Deleuze and Guattari, *A Thousand Plateaus*.

55   In Asia, the native range of rhesus macaques, a marked breeding season takes place from October to December, with the majority of births typically coinciding with the period of highest food abundance. Our research occurred during the middle of mating season for the Silver River population, between October and February, with the first birth of the year observed on April 2 (Tiffany Wade, unpublished data). Mehlman et al., "CSF 5-HIAA, Testosterone, and Sociosexual Behaviors"; Qu et al., "Rhesus Monkeys (*Macaca mulatta*) in the Taihang Mountains"; Wolfe and Peters, "History of the Freeranging Rhesus Monkeys."

56   Compare Candea, "We Both Wait Together."

57   Compare Candea, "We Both Wait Together."

58   Warkentin, "Interspecies Etiquette."

59   Fuentes et al., "Characterizing Human-Macaque Interactions in Singapore."

60   Fuentes, "Naturalcultural Encounters in Bali"; Fuentes, Shaw, and Cortes, "Qualitative Assessment of Macaque Tourist Sites."

61   Fuentes and Gamerl, "Disproportionate Participation by Age/Sex Classes"; Fuentes, Shaw, and Cortes, "Qualitative Assessment of Macaque Tourist Sites."

62   Fuentes, "Naturalcultural Encounters in Bali," 613.

63   Montague et al., "Issues and Options Related to Management of Silver Springs Rhesus Macaques," 7.

64   Compare Taussig, *Shamanism, Colonialism, and the Wild Man*; Mullin, "Mirrors and Windows."

65   Compare Lawrence, *Rodeo*; Ogden, *Swamplife*.

66   Rose, *Reports from a Wild Country*.

67   Compare Parreñas, "Producing Affect."

68   Mark Hollis, "Senate Committee Scuttles Birth Control Bill," *Ocala Star-Banner*, March 23, 1993, B1.

69   Fortwangler, "A Place for the Donkey," 214.

70. Scott Cheslak, "rhesus," e-mail, March 2, 2011.

71. Fred Hiers, "Catching, Selling, Silver River Monkeys Is Lucrative," *Gainesville*

Sun, January 5, 2012, http://www.gainesville.com; Scott Cheslak, phone interview with author, July 12, 2015.

72 Montague et al., "Issues and Options Related to Management of Silver Springs Rhesus Macaques," 6.

73 Van Dooren, "Invasive Species in Penguin Worlds," 290.

74 Craig Pittman, "State Park Will No Longer Allow Trapper to Catch Wild Monkeys for Labs," *Tampa Bay Times*, October 25, 2013, http://www.tampa bay.com.

75 Eloísa Ruano Gonzáles, "Silver Springs Monkeys Spreading as Population Grows," *Orlando Sentinel*, January 8, 2012, http://articles.orlandosentinel .com.

76 Comaroff and Comaroff, "Naturing the Nation," 251.

77 Helmreich, *Alien Ocean*, 151.

78 "Rhesus Monkeys (Macaques) in Florida," YouTube, uploaded by The-body2u, January 16, 2012, https://www.youtube.com/watch?v =PtSoFnVNWXA.

79 See Goetsch et al., "Chronic Over Browsing and Biodiversity Collapse."

80 See Weisser and Siemann, *Insects and Ecosystem Function*.

81 Lowe, *Wild Profusion*, 49.

82 Fortwangler, "A Place for the Donkey," 218.

83 Compare Haraway, "Speculative Fabulations for Technoculture's Generations," 243.

84 The management plan noted the presence of the site, but only in relation to archaeological remains from a much earlier time period: "The Paradise Park site contains successive layers of occupation, which date from Paleo-Indian times (12,000–10,000 B.P.) through the Woodland tradition. . . . Its condition assessment is good. Therefore, park development and visitor use should avoid the general area of the site." State of Florida, "Silver River State Park."

85 Graham was followed by Wendy Adams, a cultural historian who was teaching at a local community college. Adams reinforced the message about the importance of acknowledging Paradise Park and also encouraged state officials not just to concern themselves with the "authentic" Florida. "Entrepreneurs have long been coming to Florida with crazy ideas," said Adams. "Florida was a fantasy landscape, a place where people could use their imagination—stick a plaster dinosaur next to the highway and ask people to come and stop and take a look. I just want to make sure you preserve that part of Florida too, what you might refer to as 'kitschy Florida' or 'tacky Florida.' It shouldn't be erased." Adams's proposals to preserve Florida's kitsch were met with wry smiles and raised eyebrows.

86 Marian Rizzo, "Animals from Silver Springs Dispersed as Far as Arizona," *Ocala Star-Banner*, April 12, 2014, http://www.ocala.com.

87 Pittman, "State Park Will No Longer Allow Trapper."

88 Rose, *Reports from a Wild Country*.

89 Compare Collard, "Putting Animals Back Together."
90 Fortwangler, "A Place for the Donkey," 210.
91 Haraway, *When Species Meet*, 46, 65.
92 Deleuze and Guattari, *A Thousand Plateaus*, 241.
93 Cassidy and Mullin, *Where the Wild Things Are Now*.
94 Compare Collard, "Putting Animals Back Together."
95 Lowe, *Wild Profusion*.

EIGHT. MULTISPECIES FAMILIES

1 Herrel and van der Meijden, "An Analysis of the Live Reptile and Amphibian Trade," 106; compare Standish et al., "Concerns about Novel Ecosystems," 298.
2 The World Wildlife Fund is more cautious and circumspect, writing, "As a guideline, traffic has calculated that wildlife products worth about 160 US billion dollars were imported around the globe each year in the early 1990s. In addition to this, there is a large and profitable illegal wildlife trade, but because it is conducted covertly no-one can judge with any accuracy what this may be worth." "Unsustainable and Illegal Wildlife Trade," WWF, accessed September 4, 2014, http://wwf.panda.org. "Stopping Illegal Wildlife Trade," EcoHealth Alliance, accessed September 3, 2014, http://www.ecohealthalliance.org. See also Jaclin, "In the (Bleary) Eye of the Tiger."
3 Many thanks are due to Irus Braverman for helping me think about animals in relation to the law and for inviting me to present a draft of this chapter at her "More Than Human Legalities" conference in September 2014 at SUNY Buffalo. Krithika Srinivasan's comments as a discussant were invaluable in helping me sharpen my arguments and muster my ethnographic evidence. See Braverman, *Zooland*; Srinivasan, "Caring for the Collective." See also Donaldson and Kymlica, *Zoopolis*, 14.
4 Shir-Vertesh, "'Flexible Personhood.'"
5 Life itself has become a multispecies spectacle in the age of biotechnology. Hidden laborers—multiple species of animals, plants, and microbes—sustain the life of humans and creatures we love. When life forms become commodities, their value is a product of social and ecological divisions of labor. Kirksey, Costelloe-Kuehn, and Sagan, "Life in the Age of Biotechnology," 185.
6 Haraway, *When Species Meet*, 45–47.
7 See also Cheah, "Spectral Nationality"; Shukin, "Animal Capital."
8 See also Mullin, "Mirrors and Windows"; Herzog, *Some We Love, Some We Hate, Some We Eat*.
9 Shir-Vertesh, "'Flexible Personhood.'"
10 Harvey, *The Condition of Postmodernity*, 124.
11 Harvey, "Flexible Accumulation through Urbanization," 261.
12 Siskind excludes relations that are not solely human from the concept of

kinship. "The recent use of terms such as 'lineage' or 'kinship' for non-human primates," Siskind contends, "adds confusion rather than clarity to these concepts. Certainly analogues can be found for all human behavior, but what is undertaken here is to analyze kinship as a human, symbolic category" ("Kinship and Mode of Production," 860, 870). Rather than adding further confusion with my notion of multispecies families, I follow Stefan Helmreich, who has used organic practices (especially lateral gene transfer) as an opportunity to rethink cultural categories, including kinship, sketching "the rise of new kinships and biopolitics organized less around practices of 'sex' than politics of 'transfer'" ("Trees and Seas of Information," 340).

13 "Illegal Wildlife Trade," WWF, accessed September 18, 2014, http://www .worldwildlife.org.

14 Shukin, *Animal Capital*, 225–26.

15 Here I am in dialogue with Deborah Bird Rose's notion of "double death." See Rose, *Wild Dog Dreaming*. Also see Franklin, "Ethical Biocapital."

16 Here I am in dialogue with Geoffrey Bowker and Susan Leigh Star's book, *Sorting Things Out: Classification and Its Consequences*. Writing of physicians, their interpretation of the latest research, and red tape, they describe torque as "a twisting of time lines that pull at each other, and bend or twist both patient biography and the process of metrication. When all are aligned, there is no sense of torque or stress." Bowker and Star, *Sorting Things Out*, 27.

17 Ahmed, *The Promise of Happiness*, 22, 41.

18 Compare Jaclin, "In the (Bleary) Eye of the Tiger."

19 Caroline Elliot is a composite character—pet names and species have been changed to protect the anonymity of sources. Aside from these names and geographical locations, all dialogue has been reproduced verbatim. Described events and interactions, which occurred in multiple households, are reported as they happened to preserve their ethnographic and biological fidelity.

20 Schneider, *A Critique of the Study of Kinship*, 187–88.

21 Feeley-Harnick, "The Ethnography of Creation," 54.

22 See Franklin, "Biologization Revisited," 305; van Dooren, *Flight Ways*, chapter 4.

23 AvianWeb, "Hahn's, Noble or Red-Shouldered (Mini) Macaws," Beauty of Birds, accessed September 18, 2014, http://beautyofbirds.com/hahnsmacaw .html.

24 Van Dooren, *Flight Ways*, 101–2.

25 Compare Pratt, *Imperial Eyes*, 6; Clifford, *Routes*, 204.

26 Franklin, *Biological Relatives*, 15.

27 Ritvo, *The Animal Estate*, 11.

28 Franklin, "Ethical Biocapital," 102.

29 Compare Marks, "We're Going to Tell These People Who They Really Are," 355.

30 Cyborgs also "take pleasure in the confusion of boundaries" and make arguments "for responsibility in their construction." Haraway, "Manifesto for
Cyborgs," 118.

31 Berlant, *Cruel Optimism*, 48.

32 She showed me the fifth edition of Peterson's *Field Guide to Birds of Eastern
and Central North America*. This book lists fourteen species of parrots and
parakeets that could be found in Florida as of 2002 (*Peterson Field Guide to
Birds of Eastern and Central North America*, 198–99). As of 2014, Florida's
Fish and Wildlife Conservation Commission lists seventy-two introduced
species in the family Psittacidae. In addition, the website reports the presence of myriad other exotic birds: herons, egrets, ibis, spoonbills, vultures,
flamingos, ducks, teal, swans, geese, hawks, kestrels, grouse, quail, jungle-
fowl, cranes, swamphens, doves, and toucans, among many others. A total
of 196 bird species have been introduced to Florida, according to the Fish
and Wildlife Conservation Commission. "NonNative Birds," Florida Fish
and Wildlife Conservation Commission, accessed August 27, 2014, http://
myfwc.com/wildlifehabitats/nonnatives/birds/.

33 For more on contact zones, see Pratt, *Imperial Eyes*, 4–6; Clifford, *Routes*,
204; Sundberg, "Conservation Encounters"; Haraway, *When Species Meet*,
216.

34 Zeb Carruthers is a pseudonym. Details have been changed to preserve his
anonymity.

35 Donna Haraway has argued that some animals are not just "good to think"
or "good to eat" but also beings that are good "to live with." Haraway, *When
Species Meet*; Kirksey, Schuetze, and Helmreich, "Introduction," 3.

36 In 1993, Kluge rechristened this species *Morelia viridis*. Since hobbyists who
keep these snakes still refer to them as "chondros," I have used this older
species name throughout.

37 Foucault, "Right of Death and Power over Life."

38 Maxwell, *The More Complete Chondro*, Kindle ed., location 4739–73.

39 For a book-length account of this occupation, see Kirksey, *Freedom in Entangled Worlds*.

40 Kirksey, *Freedom in Entangled Worlds*, chapter 1.

41 Sarah Franklin insists that we understand kinship as a technology that acti-
vates reproductive substance, controlling and redirecting possible biological
relations. By creatively mixing and matching distinct pedigrees, Sam was
pushing biological potentials for reproduction in new directions. Producing
successful offspring would be the marker of success—transforming Sam
from someone who was embedded within a distinguished kinship network
to a patrilineal progenitor with the power to establish his own lines. Franklin, *Biological Relatives*, 152–53.

42 Here I am torquing the words of Donna Haraway, who presses for "taking
animals seriously as workers without the comforts of humanist frameworks
for people or animals." Haraway, *When Species Meet*, 73.

43 Debord, *The Society of the Spectacle*, 5.

44 Debord, *The Society of the Spectacle*, 8.

45 Rabinow and Rose, "Biopower Today," 197.

46 Ahmed, *The Promise of Happiness*, 42.

47 "*Python regius*," Wikipedia, accessed August 28, 2014, http://en.wikipedia.org.

48 Compare Bureaud, "The Ethics and Aesthetics of Biological Art," 39; Zurr, "Complicating Notions of Life," 402.

49 Anthony Caponetto Reptiles, "Ball Python Care Sheet," accessed September 17, 2014, http://www.acreptiles.com.

50 Haraway, *When Species Meet*, 45–47.

51 Williamson and Fitter, "The Varying Success of Invaders," 1651–62.

52 Donna Haraway suggests that the command "Thou shalt not kill" should be torqued into a different mandate: "Thou shalt not make killable." "The problem is to learn to live responsibly within the multiplicitous necessity and labor of killing," Haraway insists, "so as to be in the open, in the quest of the capacity to respond in relentless historical, nonteleological, multi-species contingency" (*When Species Meet*, 80). Casting my lot with Jenny, as she recounted the contingencies of entangled epidemiological and ecological problems, I began to tussle with Haraway's mandate. In the case of the Gambian pouched rat, I concluded that state agents made the right decision in making a hazardous animal killable. Rats and snakes do not claim the same spokespeople as more charismatic animals, like monkeys. But partial perspectives arguably always guide consequential cuts in multispecies worlds. Rather than accepting the wild anything-goes logic of relativism, it is necessary to make high-stakes and potentially arbitrary judgments about harm and harmlessness. Situated knowledges are guiding many decisions about making kinds of beings killable in emergent ecosystems. Paxson, "Microbiopolitics," 115–21; compare Kirksey, Costelloe-Kuehn, and Sagan, "Life in the Age of Biotechnology," 200–201.

53 Jenny Novak told a reporter that Floridians can kill and remove pythons on private property if they do it legally and humanely. "The best way to report a snake is by taking a picture and calling the hotline 888-IveGot1," wrote Aleese Kopf for the *Palm Beach Post*. "Live snakes allow experts to do important research, including tracking their movement." Aleese Kopf, "Python Patrol Participants Learn Basics of Catching Invasive Snakes," *Palm Beach Post*, July 11, 2014, http://www.palmbeachpost.com.

54 Rose, *Wild Dog Dreaming*.

55 Ahmed, *The Promise of Happiness*, 92.

56 Ahmed, *The Promise of Happiness*, 22.

57 We developed a number of theories to explain this predilection. Perhaps he preferred a higher-pitched female voice to a lower-pitched male voice. My partner also occasionally played him YouTube videos of other green tree frogs calling. So perhaps he associated her with other sounds of frog sociality.

58 Candea, "I Fell in Love with Carlos the Meerkat," 249.

59 Steve had been orphaned from his ecosystem by the agro-industrial supply chain—transported from northern Australia to the cooler climes of Sydney in a box of bananas. After Steve began hopping around a grocery store, someone called the Frog and Tadpole Study Group hotline. This conservation organization works to place orphaned amphibians in the homes of human caretakers.

60 Stuart et al., "Status and Trends of Amphibian Declines."

61 For an account of animals rescued from smuggling rings, see Collard, "Putting Animals Back Together, Taking Commodities Apart."

62 Collard, "Putting Animals Back Together, Taking Commodities Apart."

63 Ahmed, *The Promise of Happiness*, 31; Shukin, *Animal Capital*, 225–26.

64 Brent L. Brock, "Challenges of Integrating the Private Sector in Amphibian Conservation Breeding Programs," *Leaf Litter* 2, no. 1 (2008): 17, http://www.treewalkers.org/leaf-litter-magazine.

65 A receptionist at Josh's Frogs told me that all of the *Phyllobates vittatus* for sale in September 2014 were captive bred in their own facilities. See also "*Phyllobates vittatus*," Josh's Frogs, accessed September 18, 2014, http://www.joshsfrogs.com.

66 "*Phyllobates vittatus*," IUCN Red List, accessed September 18, 2014, http://www.iucnredlist.org.

67 As of September 2014, this frog was not listed under the U.S. Fish and Wildlife Endangered Species Act. While there are no federal restrictions on the sale of *Phyllobates vittatus*, state laws also govern amphibians in the pet trade. "Listed Animals: Environmental Conservation Online System," U.S. Fish and Wildlife Service, accessed September 18, 2014, http://ecos.fws.gov.

68 Many frogs previously classed as *Dendrobates* were moved to the genus *Ranitomeya* in 2006. As the objects of their conservation initiatives morphed, the taxon management plans of Tree Walkers were adjusted to embrace these new kinds of beings. "Taxon Management Plans," Tree Walkers International, accessed September 18, 2014, http://www.treewalkers.org.

69 See also Mullin, "Mirrors and Windows"; Herzog, *Some We Love, Some We Hate, Some We Eat*.

70 Brock, "Challenges of Integrating the Private Sector," 17.

71 Here I am in dialogue with Deborah Bird Rose's notion of "double death." See Deborah Bird Rose, "When All You Love Is Being Trashed," blog, September 14, 2013, http://deborahbirdrose.com. See also Rose, *Wild Dog Dreaming*.

72 "*Agalychnis annae*," IUCN Red List, accessed September 18, 2014, http://www.iucnredlist.org.

73 In the decade before the blue-sided tree frog (*Agalychnis annae*) was officially designated as endangered, the United States imported 221,960 *Agalychnis* frogs for sale in the pet trade. Most of these imports were *Agalychnis callidrayas*, the red-eyed tree frog, one of the most common pet frogs

sold in the United States. The Species Survival Network, an organization based in Washington, DC, reported that *A. annae*, *A. saltator*, and *A. spurrelli* were only occasionally sold in the international market. Species Survival Network, "Tree Frogs, *Agalychnis* spp.: SSN Fact Sheet, XV Conference of the Parties, 13–25 March, 2010," accessed September 18, 2014, http://www.ssn.org.

74 Michael Ready, Danté Fenolio, and Paige Howorth, "Notes on the Captive Care and Reproduction of the Yellow-Eyed Leaf Frog (*Agalychnis annae*)," *Leaf Litter* 3, no. 2, 14–25, http://www.treewalkers.org/leaf-litter-magazine.

75 Stengers, *Cosmopolitics II*, 35–36; Amin and Thrift, *Arts of the Political*, 132.

76 Adaptations enabling frogs to have a relatively happy existence alongside the constant presence of potential predators, like humans, might not serve them well in a world where other kinds of primates might enjoy an amphibian as a tasty snack. Brock, "Challenges of Integrating the Private Sector," 17.

77 Berlant, *Cruel Optimism*, 48.

78 "*Agalychnis annae*," IUCN Red List.

79 Marder, *Plant Thinking*, 85; Deleuze and Guattari, *A Thousand Plateaus*, 12.

NINE. PARASITES OF CAPITALISM

1   Austin, *How to Do Things with Words*, 40.

2   Dawkins and Krebs, "Animal Signals"; Davies and Halliday, "Deep Croaks and Fighting Assessment in Toads," 683–85.

3   Ahmed, *The Promise of Happiness*, 22.

4   Latour, *Politics of Nature*, 67, 69, 231–32, 249.

5   For a critique of Latour's parliament, see Sloterdijk, "Atmospheric Politics," 949.

6   Anzaldua, *Borderlands / La Frontera*.

7   For an account of architectures of apartheid, see Bowker and Star, *Sorting Things Out*.

8   For the concept of "oblique powers," see García Canclini, *Hybrid Cultures*.

9   Stepan, *Picturing Tropical Nature*, 11.

10 Quoted in Angert et al., "Cattle, Cattails, and Saltwater."

11 Sanchez, Rodriguez, and Salas, *Distribución, ciclos reproductivos y aspectos ecológicos de aves acuáticas*; Burnidge, "Cattle and the Management of Freshwater Neotropical Wetlands in Palo Verde National Park, Guanacaste, Costa Rica."

12 Gill, "A Naturalist's Guide to the OTS Palo Verde Field Station."

13 McCoy and Rodriguez, "Cattail (*Typha domingensis*) Eradication Methods," 471; Florencia Trama, personal communication, November 2010.

14 "Cattle," *Online Etymology Dictionary*, accessed October 11, 2012, http://www.etymonline.com.

15 Mitchell, *Rule of Experts*, 30. Also see Cheah, "Spectral Nationality"; Shukin, "Animal Capital."

16 Anthropologists once studied "vanishing worlds." But in the last two decades, scholarly discussions have turned toward culture defined not as "tradition" but as the world-making networks, geographies, innovations, meanings, and assemblages that are carrying us into the future. See "Emerging
Worlds," Department of Anthropology at UC Santa Cruz, accessed October
11, 2012, http://anthro.ucsc.edu.

17 Crosby, *The Columbian Exchange*, 75.

18 See also Kockelman, "Enemies, Parasites, and Noise," 406–21.

19 Stengers, *Cosmopolitics II*, 390.

20 Stengers, *Cosmopolitics I*, 74.

21 Duncan and Markoff, "Civilization and Barbarism," 34.

22 Netz, *The Cutting Edge*, viii.

23 Marx, *Capital*, 170; Derrida, *Specters of Marx*, xx; Cheah, "Spectral Nationality," 227.

24 Calvo-Alvarado et al., "Deforestation and Forest Restoration," 931–40.

25 Duncan and Markoff, "Civilization and Barbarism," 35.

26 Mitchell, *Rule of Experts*, 30; Youatt, "Counting Species."

27 Haraway, *The Companion Species Manifesto*, 36; see also Haraway, *When Species Meet*, 74, 164.

28 Daunbenmire, "Ecology of *Hyparrhenia rufa*," 11–23; Gerhardt, "Tree Seedling Development in Tropical Dry Abandoned Pasture," 95–102.

29 Kearns and Inouye, "Pollinators, Flowering Plants, and Conservation Biology," 297–307.

30 Deleuze and Guattari, *A Thousand Plateaus*, 6, 9.

31 Daunbenmire, "Ecology of *Hyparrhenia rufa*," 15.

32 Wallerstein, *The Decline of American Power*.

33 Calvo-Alvarado et al., "Deforestation and Forest Restoration," 934.

34 Gill, "A Naturalist's Guide to the OTS Palo Verde Field Station," 9.

35 "Species of Biocapital," an authoritative review by Stefan Helmreich, explores literature suggesting "that in the age of biotechnology, when the
substances and promises of biological materials, particularly stem cells and
genomes, are increasingly inserted into projects of product-making and
profit-seeking, we are witnessing the rise of a novel kind of capital" (463).
This work by Helmreich, and others reviewed by him, influences my understanding of the multiple companion species of capital, whose bodies have
been possessed and abandoned by this fickle spirit.

36 Gill, "A Naturalist's Guide to the OTS Palo Verde Field Station"; McCoy and
Rodriguez, "Cattail (*Typha domingensis*) Eradication Methods."

37 Gill, "A Naturalist's Guide to the OTS Palo Verde Field Station," 9.

38 See also Haraway, *When Species Meet*, 80.

39 Huge aqueducts built in the early 1980s, to bring water to rice parcels in
nearby villages, dramatically altered the hydrology of the region. After the
failed attempt to control cattails with cattle, the managers of Palo Verde
began to speculate that these hydrological changes were driving the cattail

growth. McCoy and Rodriguez, "Cattail (*Typha domingensis*) Eradication Methods"; see also Calo, "Influencia de las Actividades de la Cuenca del Rio Tempisque."

40 McCoy, "Seasonal, Freshwater Marshes in the Tropics," 352–53.

41 McCoy and Rodriguez, "Cattail (*Typha domingensis*) Eradication Methods."

42 By 2006 Galo was charging some $70 USD per hectare to crush cattails in Palo Verde, but this was hardly enough to cover his expenses. I obtained this figure by cross-checking the reports from Galo (Edgardo Aragon) with Florencia Trama and Ulises Chavarria. Over five hundred hectares of cattails have been crushed with tractors in Palo Verde since 1989, with a total approximate cost of $39,000.

43 Gonzales secured funds for creating and monitoring duck habitat from Costa Rica's Environment Ministry, Idea Wild, the U.S. Fish and Wildlife Service, and the Costa Rica–USA Foundation, among other sources. Trama, "Manejo Activo y Restauracion del Humedal Palo Verde."

44 A multinational regime of biopower was impinging upon local articulations of sovereign power. Representatives of the Costa Rican government were exercising their power to identify "bare life," singling out certain species of creatures as killable, in conversation with foreign agents who were exporting their own brand of sovereignty abroad. If biopower has become, in the words of Rafi Youatt, "a form of ecologically distributed power that involves interventions in human and nonhuman lives," then some of Foucault's original insights ring as true as ever: "It is not that life has been totally integrated into techniques that govern and administer it," he wrote; "it constantly escapes them." Youatt, "Counting Species," 409; Foucault, *History of Sexuality*, 143.

45 "About Ducks Unlimited," Ducks Unlimited, accessed October 11, 2012, http://www.ducks.org.

46 "Wetlands and Grassland Habitat," Ducks Unlimited, accessed October 11, 2012, http://www.ducks.org.

47 Wesley, "Socio-duckonomics," 136–42.

48 Wesley, "Socio-duckonomics," 137.

49 In contrast, ecotourists who regard certain bird species as spectacular often fail to consider the social and ecological relations that produce the charismatic creatures of interest. Vivanco, "Spectacular Quetzals."

50 U.S. Department of the Interior et al., *National Survey of Fishing, Hunting, and Wildlife-Associated Recreation*.

51 Also see Foucault, "Right of Death and Power over Life," 258–72; Agamben, *Homo Sacer*; Youatt, "Counting Species."

52 Michael Furtman, "Retooling Minnesota's Duck Factory," *Minnesota Conservation Volunteer*, September–October 2009, 10–19, http://www.dnr.state.mn.us.

53 Haraway, *When Species Meet*, 296.

54 Sawyer, *Our Sport*.

55 Cartier, *Getting the Most Out of Modern Waterfowling*, 4.

56 Tsing and Pollman, "Global Futures," 109.

57 Gill, "A Naturalist's Guide to the OTS Palo Verde Field Station," 9.

58 Deleuze and Guattari, *A Thousand Plateaus*, 456.

59 Anzaldua, *Borderlands / La Frontera*, 1.

60 Botero, "Ecology of Blue-Winged Teal," 561–65.

61 Brown, "Microparasites and Macroparasites," 161, 168.

62 Marder, *Plant Thinking*, 81.

63 Brown, "Microparasites and Macroparasites," 161.

64 Rizo, "Monitoreo de los Arrozales del Proyecto Tamarindo."

65 Rizo, "Monitoreo de los Arrozales del Proyecto Tamarindo," 4.

66 Villalobos-Brenes, Castillo, and Morales, "Efectos Toxicos y Alteracion Endocrina en la Ictiofauna."

67 Ortiz, *Cuban Counterpoint*, 139.

68 Janzen and Martin, "Neotropical Anachronisms," 19.

69 For more on hopes pinned on a despised figure, see Kirksey, *Freedom in Entangled Worlds*, 25–28.

70 See also Tsing, "Blasted Landscapes"; Kirksey, Shapiro, and Brodine, "Hope in Blasted Landscapes."

TEN. POSSIBLE FUTURES

1 Vivanco, *Green Encounters*, 5.

2 Burlingame, "Conservation in the Monteverde Zone."

3 "Introducing Monteverde and Santa Elena," Lonely Planet, accessed September 19, 2014, http://www.lonelyplanet.com.

4 Honey, *Ecotourism and Sustainable Development*, 5.

5 Vivanco, *Green Encounters*, 16.

6 Debord, *The Society of the Spectacle*, 8, 5; Vivanco, "Spectacular Quetzals, Ecotourism, and Environmental Futures," 83, 90.

7 Vivanco, "Spectacular Quetzals, Ecotourism, and Environmental Futures," 84; Vivanco, *Green Encounters*, 46n6, 88–89.

8 This brilliantly colored amphibian, *Incilius periglenes*, looked much like its distant relative, the golden frog of Panama, *Atelopus zetecki*.

9 Crump, Hensley, and Clark, "Apparent Decline of the Golden Toad," 413.

10 Honey, *Ecotourism and Sustainable Development*, 3.

11 Kenneth Burke has described the use of "four master tropes" in literature: metaphor (perspective, based on the principle of similitude), metonymy (reduction, based on that of contiguity), synecdoche (representation, based on the identification of parts of a thing as belonging to a whole), and irony (dialectic, based on opposition). Four plots correspond with these tropes in historical discourse, according to Hayden White: romance, tragedy, comedy, and satire. Tales of endangered amphibians contain elements of each story line. Conservationists have written about the golden toad in a romantic

register, while they ironically may have helped bring about the extinction of the animals they loved. Against the tragic backdrop of global climate change and mass extinction, this story is ultimately a comedy—where everything turns out all right for some of the key characters. Burke, "Four Master Tropes," 503–4; White, *Figural Realism*, 11.

12　Crump, *In Search of the Golden Frog*, 157.

13　Mello, Townsend, and Filardo, "Reforestation and Restoration at the Cloud Forest School," 148.

14　Tsing, "Indigenous Voice."

15　García Canclini, *Hybrid Cultures*, 182. See also Bowker, "Biodiversity, Datadiversity."

16　Marder, *Plant Thinking*, 85.

17　Koptur, "Breeding Systems of Monteverde *Inga*," 85.

18　Nectar is secreted by glands on young leaves, which usually dry up as the leaf matures. Ants are less abundant at higher elevations, as in the Monteverde cloud forest. *Inga* nectaries at high elevations attract parasitic wasps and flies that lay eggs inside caterpillars and other insect herbivores. Larval wasps and fly maggots slowly kill these herbivores after hatching out inside their bodies and eating them from within.

19　Janzen, *Costa Rican Natural History*, 259–60; Koptur, "Interactions among *Inga*, Herbivores, Ants, and Insect Visitors," 277–78. See also Bentley, "Extrafloral Nectaries and Protection by Pugnacious Bodyguards," 407–27; Leston, "The Ant Mosaic," 311–41.

20　Latour, *Politics of Nature*, 455.

21　Quote from Lamb et al., "Rejoining Habitat Remnants," 371. Other key papers include Holl, "Effect of Shrubs on Tree Seedling Establishment in an Abandoned Tropical Pasture"; Peterson and Haines, "Early Successional Patterns," 361–69.

22　Marder, *Plant Thinking*, 40.

23　Eladio Cruz identified the plants growing on this hollow heart tree in February 2014.

24　Haber, Zuchowski, and Bello, *An Introduction to Cloud Forest Trees*, 20–21, 145.

25　In Panama the Nature Conservancy supported U.S. Department of Defense agendas by issuing a report that characterized former bombing ranges as "relatively pristine." This is discussed in chapter 3, "Hope in the Reverted Zone."

26　Burlingame, "History of the Cloud Forest School."

27　The Monteverde Conservation League was founded in 1986 "to conserve, preserve, and rehabilitate tropical ecosystems and their biodiversity." Leslie Burlingame has written a brief history of their efforts:

> Tree seedlings were produced with help from the U.S. Peace Corps and other volunteers and delivered to farms by Monteverde Conservation League

personnel, who also provided technical assistance. The farmers had to invest their own labor, which tied them to the project. They received financial incentives provided by The Netherlands through the Costa Rican Forest Service to cover expenses of planting and fencing in the form of a loan that was forgiven if the farmer cared for the trees for three years. . . . By 1994, more than 500,000 trees produced in the nurseries had been planted by 263 farmers in 320 windbreak projects. Farmers perceived that windbreaks increased milk and crop production and contributed to their wood needs and to wildlife habitat.

Burlingame, "Conservation in the Monteverde Zone," 355–56.

28 Vivanco, *Green Encounters*, 97–98.

29 For an account of animals as laborers, see Haraway, *When Species Meet*.

30 A plant specialist from the Unites States who moved to Monteverde some thirty years ago, Willow Zuchowski, is concerned that these alien earthworms will escape from compost bins to wreak havoc on the local vermifauna. "We lack even basic research here about local worms," Willow told me. "We don't yet know what is here, much less how these imported worms will change the soil community ecology." The identity of the worms in Monteverde compost bins is still an open question. Composters in California, the rumored origin point of the worms in Monteverde, commonly use *Eisenia fetida*, an animal species that is derived from European stock. According to *Worms Eat My Garbage*, a do-it-yourself guide to composting that I discovered in a Monteverde library, "Dr. Roy Hartenstein of Syracuse, New York, has calculated that eight individual [*Eisenia fetida*] worms could produce about 1500 offspring within six-months-time. . . . With the reproductive potential described, we come to the question of why these worms don't take over the world" (Appelhof, *Worms Eat My Garbage*, 28). The short answer to this very complex question is that these worms require high concentrations of nutrients. Individuals of this species usually die when they are transplanted into garden plots. Monteverde's community of expatriate biologists lacks an expert on the rhizosphere, the world of microorganisms and roots in a narrow region of surface soil. No one can authoritatively speak for nature; no one can confidently represent the interests of the multiple species who are living and dying in the dark depths of vermiculture projects.

31 These foreign worms are like *Nature's Little Helpers*, fabulous sculptures created by Patricia Piccinini that reframe the problem of caring for endangered organisms of the Australian outback. These sculptures depict genetically modified extraterrestrials who have come to earth to help wombats and birds, critters that have been pushed to the brink of extinction by people. Donna Haraway's essay "Speculative Fabulations for Technoculture's Generations" contrasts these artworks within twenty-first-century technoscientific frontier practices that are "always announcing new worlds,

proposing the novel as the solution to the old, figuring creation as radical invention and replacement, rushing toward a future that wobbles between ultimate salvation and destruction but has little truck with thick pasts or presents." Rather than proposing another frontier, the helpful aliens of Piccinini's imagination embody "something more akin to a decolonizing ethic indebted to Australian Aboriginal practices of taking care of country and accounting for generations of entangled human and nonhuman entities." Haraway, "Speculative Fabulations for Technoculture's Generations," 243.

32 Van den Akker and Soane, "Compaction," 285.

33 Kennedy and de Luna, "Rhizosphere," 399.

34 Haraway, *The Companion Species Manifesto*.

35 Appelhof, *Worms Eat My Garbage*, 74.

36 Harris, "Do Feedbacks from the Soil Biota Secure Novelty," 125.

37 Marder, *Plant Thinking*, 40.

38 Richard, "Compost," 300.

39 Kristina Lyons describes how soils became a matter of political concern in the U.S.-Colombia "War on Drugs." State scientists worked to identify productive and market-oriented soils that can be improved by human action. Putumayo farmers engage in alternative material practices with soils that are formed by entangled life-propagating relations. Lyons, "Soil Science, Development, and the 'Elusive Nature' of Colombia's Amazonian Plains," 212.

40 Appelhof, *Worms Eat My Garbage*, 66.

41 Compare Agamben, *The Open*, 2; Vivanco, *Green Encounters*.

42 Paxson, "Microbiopolitics."

43 Compare with Balakrishnan and Schmitt, *The Enemy*; Buck-Morss, *Dreamworld and Catastrophe*; Kirksey, *Freedom in Entangled Worlds*, 177.

44 Rose, *Wild Dog Dreaming*, 18–19.

45 For more on "twisting death back into life" and avoiding the forms of eradication that result in "double death," see Rose, *Wild Dog Dreaming*; van Dooren, "Vultures and Their People in India." For more on the necessity of killing, see Haraway, *When Species Meet*, 80, 105–6.

46 Patricia Townsend, who conducted experimental manipulations on the Monteverde Cloud Forest School lands for her PhD dissertation, had unexpected findings. Rather than plant seedlings germinated in a greenhouse, Townsend's experiment involved planting some eight thousand seeds in the abandoned pasture. She removed star grass from some plots with repeated chopping and "solarized" treatment. But she found that "both grass reduction treatments did not improve germination. Instead controls, which were shady and damp [with the star grass intact], had the highest germination" for four of the tree species included in her study. "Although presence of grass may be beneficial to germination," Townsend wrote, "grass may compete with seedlings once they are taller. . . . In my study, many seedlings in the controls were smothered in the dense grass after 1 year and as a

result either rotted or became bent over at the top." Townsend, "Conservation and Restoration for a Changing Climate," 10–11.

47 Marder, *Plant Thinking*, 131, 138.

48 Lévinas, *Otherwise Than Being*, 4.

49 Haber, Zuchowski, and Bello, *An Introduction to Cloud Forest Trees*, 93.

50 Powell justification attached to memo by the Nature Conservancy's Randy Curtis, March 16, 1992. Quoted in Burlingame, "History of the Cloud Forest School."

51 Deborah Hamilton used the research of Carlos Guindon, who studied trees in forest fragments of Costa Rica's Pacific Slope for his doctoral thesis at Yale, to guide her vision of "natural" abundance. She took Guindon's data about percentages of different tree species in forest fragments and then sought to produce roughly the same percentages of trees in her nursery. A data sheet that she shared with me included few numbers from Guindon to guide the production of the Lauraceae species she regarded as "super quetzal and bellbird trees." As of 2007, Deborah Hamilton had produced 3,004 *Ocotea whiteii*, 2,374 *Ocotea monteverdensis*, and 2,174 *Inga punctata* saplings, which represented 7.6 percent, 6.0 percent, and 5.5 percent of the total trees she grew. Guindon's data for these species were not included in her spreadsheet. By 2007 she had produced 50,356 saplings, representing sixty-seven species in twenty-eight families.

52 Donna Haraway first posed these questions: Haraway, *Modest_Witness*, 104.

53 Janzen, *Costa Rican Natural History*, 133, 200–201.

54 Clark, "Abolishing Virginity."

55 Milton's understanding of care is kindred to recent feminist scholarship on the subject. Practices of care are often too self-centered, according to Maria Puig de la Bellacasa. Care is often regarded as "everything that we do to maintain, continue and repair our world," which is constituted by "our bodies, our selves, and our environment." Building on the work of Donna Haraway, Puig suggests that we should decenter our world by caring for other kinds of critters. "To care about something, or for somebody," writes Puig, "is inevitably to create relation." Puig de la Bellacasa, "'Nothing Comes without Its World,'" 198.

56 Tsing, "Contaminated Diversity in 'Slow Disturbance,'" 97.

57 The thirty-eight trees identified by Eladio included three known unknowns—three species that were known to local ecologists but that had not yet been described and formally named by botanists. The species he identified were *Myrsine coriacea* (Myrsinaceae), *Ocotea floribunda* (Lauraceae), *Daphnopsis americana* (Thymelaeaceae), *Solanum* sp. 1 (Solanaceae), *Cinnamomum costaricanum* (Lauraceae), *Sideroxylon puertoricensis* (Sapotaceae), *Sapium glandulosum* (Euphorbiaceae), *Zygia longifolia* (Fabaceae/Mimosoideae), *Erythrina* sp (Fabaceae/Papilionoideae), *Sideroxylon persimile* (Sapotaceae), *Myrcianthes* sp. black (Myrtaceae), *Beilschmiedia brenesii* (Lauraceae), *Cedrela tonduzii* (Meliaceae), *Palicourea padifolia* (Rubiaceae),

Diospyros hartmanniana (Ebenaceae), *Montanoa guatemalensis* (Asteraceae), *Casimiroa edulis* (Rutaceae), *Ocotea tenera* (Lauraceae), *Saurauia montana* (Actinidiaceae), *Clusia* sp. A (Clusiaceae), *Clusia* sp. B (Clusiaceae), *Clusia* sp. C (Clusiaceae), *Sapium glandulosum* (Euphorbiaceae), *Cestrum panamense* (Solanaceae), *Beilschmiedia alloiophylla* (Lauraceae), *Hampea appendiculata* (Malvaceae), *Nectandra membranacea* (Lauraceae), *Ardisia compressa* (Myrsinaceae), *Persea* sp. A (Lauraceae), *Trichilia havanensis* (Meliaceae), *Ocotea tonduzii* (Lauraceae), *Hasseltia floribunda* (Salicaceae), *Matayba oppositifolia* (Sapindaceae), *Viburnum costaricanum* (Caprifoliaceae), *Citharexylum costaricensis* (Verbenaceae), *Ficus velutina* (Moraceae), *Xylosma intermedia* (Salicaceae, Flacourtiaceae), and *Zanthoxylum melanostictum* (Rutaceae).

58 Nadkarni and Wheelwright, *Monteverde*, 41; Burlingame, "History of the Cloud Forest School."

59 William A. Haber, "Plants of Monteverde: Identification and Other Resources," UMass, Boston, Department of Computer Science, accessed April 27, 2015, http://www.cs.umb.edu.

60 See also Melson, "Prehistoric Eruptions of Arenal Volcano," 35–59.

61 Sheets et al., "Prehistory and Volcanism in the Arenal Area," 445, 462.

62 "Ruins are now our gardens," writes Anna Tsing. "Degraded ('blasted') landscapes produce our livelihoods." These "rubble ecologies," to borrow a phrase from Bettina Stoetzer, involve the flourishing of weeds and displaced peoples amid collapse. Stoetzer, *At the Forest Edges of the City*; Tsing, "Blasted Landscapes."

63 Compare with Hardt and Negri, *Multitude*; Thacker, "Networks, Swarms, Multitudes."

64 Compare Geoghegan, "An Anti-humanist Utopia?," 49–51.

65 Ahmed, *The Promise of Happiness*, 31.

CONCLUSION

1 For comparison, see Deleuze and Guattari, *A Thousand Plateaus*, 36.

2 For a description of niches emerging at the human-nonhuman interface, see Fuentes, "Naturalcultural Encounters in Bali," 604; see also Popielarz and Neal, "The Niche as a Theoretical Tool."

3 Chrulew, "Managing Love and Death at the Zoo."

4 Haraway, "Speculative Fabulations for Technoculture's Generations." See also Heatherington, "From Ecocide to Genetic Rescue"; van Dooren, *Flight Ways*.

5 For more on twisting death back into life, see Rose, *Wild Dog Dreaming*; van Dooren, "Vultures and Their People in India." For more on the impossibility of a final solution, see Haraway, *When Species Meet*, 106; van Dooren, "Invasive Species in Penguin Worlds." On dwelling with the ethical difficulties of killing in conservation, see van Dooren, *Flight Ways*, chapter 4.

6 "There is no way to eat and not to kill," writes Donna Haraway in *When Spe-*

cies Meet, "no way to pretend innocence and transcendence or a final peace. . . . The practice of regard and response has no preset limits, but giving up human exceptionalism has consequences that require one to know more at the end of the day than at the beginning and to cast oneself with some ways of life and not others in the never settled biopolitics of entangled species" (295). Pushing these insights beyond the realm of foodways, it is also clear that living in emergent ecologies means that we must cast our lots with some ways of life and not others.

7 This paragraph is in close dialogue with an anonymous pamphlet called *Desert*, written by "a nature loving anarchist," which suggests that many activists, anarchists, and environmentalists are haunted by a realization: "The world will not be 'saved.' Global anarchist revolution is not going to happen. Global climate change is now unstoppable. We are not going to see the worldwide end to civilization/capitalism/patriarchy/authority. . . . This realisation hurts people." Anonymous, *Desert*, 7–8, Anarchist Library, accessed September 19, 2014, http://theanarchistlibrary.org. Thanks are due to Anna Tsing for pointing me toward this pamphlet and for also showing me how to garden in ruins.

8 Tsing, "Blasted Landscapes"; Kirksey, Shapiro, and Brodine, "Hope in Blasted Landscapes"; Bolender, "R.A.W. Assmilk Soap."

9 Marder, *Plant Thinking*, 142; Paxson, "Microbiopolitics"; Raffles, *The Illustrated Insectopedia*.

BIBLIOGRAPHY

Adams King, Wendy. "Through the Looking Glass of Silver Springs: Tourism and the Politics of Vision." *Americana: The Journal of American Popular Culture* 3, no. 1 (2004).

Agamben, Giorgio. *Homo Sacer: Sovereign Power and Bare Life*. Stanford, CA: Stanford University Press, 1998.

———. *The Open: Man and Animal*. Stanford, CA: Stanford University Press, 2004.

Ahmed, Sara. *The Promise of Happiness*. Durham, NC: Duke University Press, 2010.

Altshuler, Douglas. "Novel Interactions of Non-pollinating Ants with Pollinators and Fruit Consumers in a Tropical Forest." *Oecologia* 119 (1999): 600–606.

Alvarado-Quesada, Guiselle Maria. "Conservacion de las Aves Acuaticas de Costa Rica." *BRENESIA* 66 (2006): 49–68.

Amin, Ash, and Nigel J. Thrift. *Arts of the Political: New Openings for the Left*. Durham, NC: Duke University Press, 2013.

Angert, Amy, Dan Ardia, Douglas Gill, Don Dougie, Corine Vriesendorp, Nina Brown, Heather Ewell, and Albert Owen. "Cattle, Cattails, and Saltwater: A Tale of Many Stories OR Pers. Comm. vs. Pers. Obs." In *OTS Course Book 99-1*, 71–94. San Jose, Costa Rica: Organization of Tropical Studies, 1999.

Anzaldua, Gloria. *Borderlands / La Frontera: The New Mestiza*. San Francisco: Spinsters / Aunt Lute, 1987.

Appelhof, Mary. *Worms Eat My Garbage*. Kalamazoo, MI: Flower Press, 1982.

Austin, John L. *How to Do Things with Words*. Oxford: Oxford University Press, 1965.

Baitchman, Eric J., and Alan P. Pessier. "Pathogenesis, Diagnosis, and Treatment of Amphibian Chytridiomycosis." In *Select Topics in Dermatology*, edited by P. G. Fisher, 669–86. Philadelphia: Elsevier, 2013.

Balakrishnan, Gopal, and Carl Schmitt. *The Enemy: An Intellectual Portrait of Carl Schmitt*. New York: Verso, 2000.

Barad, Karen. "Invertebrate Visions: Diffractions of the Brittlestar." In *The Multispecies Salon*, edited by E. Kirksey, 221–41. Durham, NC: Duke University Press, 2014.

———. "Meeting the Universe Halfway: Realism and Social Constructivism without Contradiction." In *Feminism, Science, and the Philosophy of Science*, edited by L. H. Nelson and J. Nelson. Dordrecht, Holland: Kluwer, 1996.

Barad, Karen Michelle. *Meeting the Universe Halfway: Quantum Physics and the Entanglement of Matter and Meaning*. Durham, NC: Duke University Press, 2007.

Bataille, Georges. *The Accursed Share: An Essay on General Economy*. New York: Zone, 1991.

Bateson, Gregory. *Steps to an Ecology of Mind: Collected Essays in Anthropology, Psychiatry, Evolution, and Epistemology*. San Francisco: Chandler, 1972.

Bateson, Mary Catherine. *Peripheral Visions: Learning along the Way*. New York: HarperCollins, 1994.

Becker, Howard. *Doing Things Together: Selected Papers*. Evanston, IL: Northwestern University Press, 1986.

Benitez Rojo, Antonio, and James E. Maraniss. *The Repeating Island: The Caribbean and the Postmodern Perspective*. Durham, NC: Duke University Press, 1996.

Benjamin, Walter. *Illuminations*. New York: Schocken, 1968.

Bennett, Jane. *Vibrant Matter: A Political Ecology of Things*. Durham, NC: Duke University Press, 2010.

Bennett, Tony. "The Exhibitionary Complex." *new formations* 4 (1988): 73–102.

Bentley, B. L. "Extrafloral Nectaries and Protection by Pugnacious Bodyguards." *Annual Review of Ecology and Systematics* 8 (1977): 407–27.

Berlant, Lauren. *Cruel Optimism*. Durham, NC: Duke University Press, 2011.

Bernstein, I. S., and C. L. Ehardt. "Intragroup Agonistic Behavior in Rhesus-Monkeys (*Macaca mulatta*)." *International Journal of Primatology* 6, no. 3 (1985): 209–26.

Bletz, Molly C., Andrew H. Loudon, Matthew H. Becker, Sara C. Bell, Douglas C. Woodhams, Kevin P. C. Minbiole, and Reid N. Harris. "Mitigating Amphibian Chytridiomycosis with Bioaugmentation: Characteristics of Effective Probiotics and Strategies for Their Selection and Use." *Ecology Letters* 16, no. 6 (2013): 807–20.

Blüthgen, Nico, Manfred Verhaagh, William Goitía, Klaus Jaffé, Wilfried Morawetz, and Wilhelm Barthlott. "How Plants Shape the Ant Community in the Amazonian Rainforest Canopy: The Key Role of Extrafloral Nectaries and Homopteran Honeydew." *Oecologia* 125 (2000): 229–40.

Bolender, Karin. "R.A.W. Assmilk Soap." In *The Multispecies Salon*, edited by E. Kirksey, 64–86. Durham, NC: Duke University Press, 2014.

Botero, Jorge Eduardo. "Ecology of Blue-Winged Teal Wintering in the Neotropics." PhD diss., University of Wisconsin, 1992.

Bourdieu, Pierre. *Distinction: A Social Critique of the Judgement of Taste*. Cambridge, MA: Harvard University Press, 1984.

Bowker, Geoffrey C. "Biodiversity, Datadiversity." *Social Studies of Science* 30, no. 5 (2000): 643–84.

———. *Memory Practices in the Sciences*. Cambridge, MA: MIT Press, 2005.

Bowker, Geoffrey C., and Susan Leigh Star. *Sorting Things Out: Classification and Its Consequences*. Cambridge, MA: MIT Press, 1999.

Bradburd, Daniel. "Adventure in the Zeitgeist, Adventures in Reality: Simmel, Tarzan, and Beyond." In *Tarzan Was an Eco-tourist: And Other Tales in the Anthropology of Adventure*, edited by L. A. Vivanco and R. J. Gordon, 43–57. New York: Berghahn, 2006.

Braun, Bruce. *The Intemperate Rainforest: Nature, Culture, and Power on Canada's West Coast*. Minneapolis: University of Minnesota Press, 2002.

Braverman, Irus. *Zooland: The Institution of Captivity*. Stanford, CA: Stanford University Press, 2013.

Breed, Michael D., P. Abel, T. J. Bleuze, and S. E. Denton. "Thievery, Home Ranges, and Nestmate Recognition in *Ectatomma ruidum*." *Oecologia* 84 (1990): 17–121.

Breed, Michael D., Terrence McGlynn, E. M. Stocker, and A. N. Klein. "Thief Workers and Variation in Nestmate Recognition in a Ponerine Ant, *Ectatomma ruidum*." *Insectes Sociaux* 46 (1999): 327–31.

Breed, Michael D., L. E. Snyder, T. L. Lynn, and J. A. Morhart. "Acquired Chemical Camouflage in a Tropical Ant." *Animal Behaviour* 44 (1992): 519–23.

Brenkman, John. "Introduction to Bataille." *New German Critique* 16 (1979): 59–63.

Brown, Peter J. "Microparasites and Macroparasites." *Current Anthropology* 2, no. 1 (1987): 155–71.

Buchanan, Brett. *Onto-Ethologies: The Animal Environments of Uexküll, Heidegger, Merleau-Ponty, and Deleuze*. Albany: State University of New York Press, 2008.

Buck-Morss, Susan. *Dreamworld and Catastrophe: The Passing of Mass Utopia in East and West*. Cambridge, MA: MIT Press, 2000.

Bureaud, Annick. "The Ethics and Aesthetics of Biological Art." *Art Press* 276 (2002): 38–39.

Burke, Kenneth. "Four Master Tropes." In *A Rhetoric of Motives*, 503–18. Berkeley: University of California Press, 1969.

Burlingame, Leslie. "Conservation in the Monteverde Zone: Contributions of Conservation Organizations." In *Monteverde: Ecology and Conservation of a Tropical Cloud Forest*, edited by N. M. Nadkarni and N. T. Wheelwright, 351–85. New York: Oxford University Press, 2000.

———. "History of the Cloud Forest School." Unpublished manuscript, Franklin and Marshall College, February 2008.

Burnidge, William S. "Cattle and the Management of Freshwater Neotropical Wetlands in Palo Verde National Park, Guanacaste, Costa Rica." Master's thesis, University of Michigan, 2000.

Calarco, Matthew. *Zoographies: The Question of the Animal from Heidegger to Derrida*. New York: Columbia University Press, 2008.

Callon, Michel. "Some Elements of a Sociology of Translation: Domestication of the Scallops and the Fishermen of St. Brieuc Bay." In *Power, Action, and Belief: A New Sociology of Knowledge?*, edited by J. Law. London: Routledge and Kegan Paul, 1986.

Calo, Julio Cesar. "Influencia de las Actividades de la Cuenca del Rio Tempisque en la Calidad del Agua." Paper presented at Taller El Agua en el Rio Tempisque:

Calidad Flujos y Conservacion, Palo Verde Biological Station, Guanacaste, Costa Rica, November 7–10, 2000.

Calvo-Alvarado, J., B. McLennan, A. Sanchez-Azofeifa, and T. Garvin. "Deforestation and Forest Restoration in Guanacaste, Costa Rica: Putting Conservation Policies in Context." *Forest Ecology and Management* 258 (2009): 931–40.

Candea, Matei. "I Fell in Love with Carlos the Meerkat: Engagement and Detachment in Human-Animal Relations." *American Ethnologist* 37, no. 2 (2010): 241–58.

———. "We Both Wait Together: Poaching Agustin Fuentes." *Kroeber Anthropological Society Papers* 100, no. 1 (2011): 148–51.

Carse, Ashley. "Nature as Infrastructure: Making and Managing the Panama Canal Watershed." *Social Studies of Science* 42, no. 4 (2012): 539–63.

Cartier, John O. *Getting the Most Out of Modern Waterfowling.* New York: St. Martin's, 1974.

Cassidy, Rebecca, and Molly Mullin. *Where the Wild Things Are Now: Domestication Reconsidered.* New York: Berg, 2007.

Cassil, D. L., J. Butler, S. B. Vinson, and D. E. Wheeler. "Cooperation during Prey Digestion between Workers and Larvae in the Ant, *Pheidole spadonia.*" *Insectes Sociaux* 52 (2005): 339–43.

Cheah, Pheng. "Spectral Nationality: The Living On (*sur-vie*) of the Postcolonial Nation in Neocolonial Globalization." *boundary 2* 26, no. 3 (1999): 225–52.

Choy, Timothy. *Ecologies of Comparison: An Ethnography of Endangerment in Hong Kong.* Durham, NC: Duke University Press, 2011.

Christen, Catherine. "At Home in the Field: Smithsonian Tropical Science Field Stations in the U.S. Panama Canal Zone and the Republic of Panama." *The Americas* 58, no. 4 (2002): 537–75.

Chrulew, Matthew. "Managing Love and Death at the Zoo: The Biopolitics of Endangered Species Preservation." *Australian Humanities Review* 50 (2011): 137–57.

Clark, David B. "Abolishing Virginity." *Journal of Tropical Ecology* 12 (1996): 735–39.

Clark, D. B., C. Guayasamin, O. Pazmino, C. Donoso, and Y. Paez de Villacis. "The Tramp Ant *Wasmannia auropunctata*: Autoecology and Effects on Ant Diversity and Distribution on Santa Cruz Island, Galapagos." *Biotropica* 14, no. 3 (1982): 196–207.

Clarke, Adele. *Disciplining Reproduction: Modernity, American Life Sciences, and "the Problems of Sex."* Berkeley: University of California Press, 1998.

Clements, Frederic. "Nature and Structure of the Climax." *Journal of Ecology* 24, no. 1 (1936): 252–84.

———. *Plant Succession: An Analysis of the Development of Vegetation.* Washington, DC: Carnegie Institution of Washington, 1916.

———. *Research Methods in Ecology.* Lincoln, NE: University Publishing Company, 1905.

Clifford, James. *Routes: Travel and Translation in the Late Twentieth Century.* Cambridge, MA: Harvard University Press, 1997.

Collard, Rosemary-Claire. "Putting Animals Back Together, Taking Commodities Apart." *Annals of the Association of American Geographers* 104, no. 1 (2014): 151–65.

Collins, Harry M., and Steven Yearley. "Epistemological Chicken." In *Science as Practice and Culture*, edited by A. Pickering. Chicago: University of Chicago Press, 1992.

Comaroff, Jean, and John Comaroff. "Naturing the Nation: Aliens, Apocalypse, and the Postcolonial State." *Social Identities* 7, no. 2 (2010): 233–65.

Connell, Joseph. "On the Role of Natural Enemies in Preventing Competitive Exclusion in Some Marine Animals and in Rain Forest Trees." In *Dynamics of Population*, edited by P. J. D. Boer and G. R. Gradwell, 298–312. Wageningen, the Netherlands: Pudoc, 1971.

Constantino, Jill. *Tortoise Soup for the Soul.* Bloomington: Indiana University Press, 2012.

Cook, O. F. "Cotton Culture in Guatemala." *Yearbook of the United States Department of Agriculture, 1904* (1905): 475–88.

———. "An Enemy of the Cotton Boll Weevil." *Science* 19, no. 492 (1904): 862–64.

———. "Professor William Morton Wheeler on the Kelep." *Science* 20, no. 514 (1904): 611–12.

———. "The Social Organization and Breeding Habits of the Cotton-Protecting Kelep of Guatemala." *Technical Series of the U.S. Department of Agriculture, Bureau of Entomology* 10 (1905): 1–55.

Cormier, Loretta A. *Kinship with Monkeys: The Guajá Foragers of Eastern Amazonia.* New York: Columbia University Press, 2003.

Crapanzano, Vincent. *Imaginative Horizons: An Essay in Literary-Philosophical Anthropology.* Chicago: University of Chicago Press, 2004.

Cronon, William. "The Trouble with Wilderness: Or, Getting Back to the Wrong Nature." *Environmental History* 1, no. 1 (1996): 7–28.

Crosby, Alfred W. *The Columbian Exchange: Biological and Cultural Consequences of 1492.* Westport, CT: Greenwood, 1972.

Crump, Martha L., Frank R. Hensley, and Kenneth L. Clark. "Apparent Decline of the Golden Toad: Underground or Extinct?" *Copeia* 2 (1992): 413–20.

Crump, Marty. *In Search of the Golden Frog.* Chicago: University of Chicago Press, 2000.

Cussins, Charis. "Ontological Choreography: Agency through Objectification in Infertility Clinics." *Social Studies of Science* 26 (1996): 575–610.

da Costa, Beatriz, and Kavita Philip. *Tactical Biopolitics: Art, Activism, and Technoscience.* Cambridge, MA: MIT Press, 2008.

Datta, S. "The Acquisition of Dominance among Free-Ranging Rhesus-Monkey Siblings." *Animal Behaviour* 36 (1988): 754–72.

Daunbenmire, R. "Ecology of *Hyparrhenia rufa* (Nees) in Derived Savanna in Northwestern Costa Rica." *Journal of Applied Ecology* 9, no. 1 (1972): 11–23.

Davidson, C., C. E. Williamson, K. Vincent, S. M. Simonich, K. S. Yip, J. M. Hero, and K. M. Kriger. "Anuran Population Declines Occur on an Elevational Gradient in the Western Hemisphere." *Herpetological Conservation and Biology* 8, no. 3 (2013): 503–18.

Davies, N. B., and T. R. Halliday. "Deep Croaks and Fighting Assessment in Toads (*Bufo bufo*)." *Nature* 274 (1978): 683–85.

Dawkins, Richard, and John R. Krebs. "Animal Signals: Information or Manipulation." In *Behavioural Ecology: An Evolutionary Approach*, edited by J. R. Krebs and N. B. Davies, 282–309. Oxford: Blackwell Scientific, 1978.

Debord, Guy. *The Society of the Spectacle*. New York: Zone Books, 1967.

de Certeau, Michel. *The Practice of Everyday Life*. Translated by S. Rendall. Berkeley: University of California Press, 1984.

de la Fuente, M. A. S., and R. J. Marquis. "The Role of Ant-Tended Extrafloral Nectaries in the Protection and Benefit of a Neotropical Rainforest Tree." *Oecologia* 118, no. 2 (1999): 192–202.

Deleuze, Gilles, and Félix Guattari. *A Thousand Plateaus: Capitalism and Schizophrenia*. London: Athlone, 1987.

Derrida, Jacques. *The Animal That Therefore I Am*. Translated by D. Wills. New York: Fordham University Press, 2008.

———. "For a Justice to Come: An Interview with Jacques Derrida." By Lieven De Cauter. Indymedia.be, April 5, 2004.

———. "Marx and Sons." In *Ghostly Demarcations: A Symposium on Jacques Derrida's Specters of Marx*, edited by M. Sprinker, 213–69. New York: Verso, 1999.

———. "Plato's Pharmacy." In *Dissemination*, 61–172. London: Athlone Press, 1981.

———. *Specters of Marx: The State of the Debt, the Work of Mourning, and the New International*. New York: Routledge, 1994.

DeVries, P. J. "Enhancement of Symbioses between Butterfly Caterpillars and Ants by Vibrational Communication." *Science* 248, no. 4959 (1990): 1104–6.

DeVries, P. J., and I. Baker. "Butterfly Exploitation of an Ant-Plant Mutualism: Adding Insult to Herbivory." *Journal of the New York Entomological Society* 97, no. 3 (1989): 332–40.

Diprose, Rosalyn. *Corporeal Generosity: On Giving with Nietzsche, Merleau-Ponty, and Levinas*. Albany: State University of New York Press, 2002.

Donaldson, Sue, and Will Kymlicka. *Zoopolis: A Political Theory of Animal Rights*. Oxford: Oxford University Press, 2011.

Drewry, George E., W. Ronald Heyer, and A. Stanley Rand. "A Functional Analysis of the Complex Call of the Frog *Physalaemus pustulosus*." *Copeia* 3 (1982): 636–45.

Duellman, William Edward, and Linda Trueb. *Biology of Amphibians*. New York: McGraw-Hill, 1986.

Duncan, Silvio R., and John Markoff. "Civilization and Barbarism: Cattle Frontiers in Latin America." In *States of Violence*, edited by F. Coronil and J. Skurski, 33–74. Ann Arbor: University of Michigan Press, 2006.

Ehrlich, Paul, and Anne Ehrlich. *Extinction: The Causes and Consequences of the Disappearance of Species*. New York: Random House, 1981.

Elden, Stuart, and Eduardo Mendieta. "Being-With as Making Worlds: The 'Second Coming' of Peter Sloterdijk." *Environment and Planning D: Society and Space* 27 (2009): 1–11.

Elepfandt, Andreas. "*Xenopus* Sensory Perception." In *The Biology of Xenopus*, edited by R. C. Tinsley and H. R. Kobel, 97–120. Oxford: Clarendon, 1996.

Eliot, Christopher. "Method and Metaphysics in Clements's and Gleason's Ecological Explanations." *Studies in History and Philosophy of Biological and Biomedical Sciences* 38, no. 1 (2007): 85–109.

Elkan, Edward R. "The Xenopus Pregnancy Test." *British Medical Journal* 4 (1938): 1253–56.

Elmore, D., and R. Eberle. "Monkey B Virus (Cercopithecine herpesvirus 1)." *Comparative Medicine* 58, no. 1 (2008): 11–21.

Encyclopaedia Britannica. *The Encyclopaedia Britannica Guide to Genetics*. London: Robinson, 2009.

Engel, Gregory A., Lisa Jones-Engel, Michael A. Schillaci, Komang Gde Suaryana, Artha Putra, Agustin Fuentes, and Richard Henkel. "Human Exposure to Herpesvirus B–Seropositive Macaques, Bali, Indonesia." *Emerging Infectious Diseases* 8, no. 8 (2002): 789–95.

Evans, B. J., D. B. Kelley, R. C. Tinsley, D. J. Melnick, and D. C. Cannatella. "A Mitochondrial DNA Phylogeny of African Clawed Frogs: Phylogeography and Implications for Polyploid Evolution." *Molecular Phylogenetics and Evolution* 33, no. 1 (2004): 197–213.

Feeley-Harnik, Gilian. "The Ethnography of Creation: Lewis Henry Morgan and the American Beaver." In *Relative Values: Reconfiguring Kinship Studies*, edited by S. Franklin and S. McKinnon, 54–84. Durham, NC: Duke University Press, 2001.

Feinberg, Harvey M., and Joseph B. Solodow. "Out of Africa." *Journal of African History* 43 (2002): 255–61.

Fjellman, Stephen M. *Vinyl Leaves: Walt Disney World and America*. Boulder, CO: Westview, 1992.

Fortun, Michael. "Mediated Speculations in the Genomics Futures Markets." *New Genetics and Society* 20, no. 2 (2001): 139–56.

Fortwangler, Crystal. "A Place for the Donkey: Natives and Aliens in the US Virgin Islands." *Landscape Research* 34, no. 2 (2009): 205–22.

Foucault, Michel. *The Birth of the Clinic: An Archaeology of Medical Perception*. Translated by A. Sheridan. London: Tavistock, 1976.

———. *Discipline and Punish: The Birth of the Prison*. Translated by A. Sheridan. London: Penguin, 1991.

———. *The History of Sexuality*, vol. 1. Translated by Robert Hurley. New York: Vintage, 1978.

———. "Nietzsche, Genealogy, History." In *Language, Counter-memory, Practice*, edited by D. F. Bouchard, 139–64. Ithaca, NY: Cornell University Press, 1977.

———. "Of Other Spaces: Utopias and Heterotopias." *Diacritics* 16, no. 1 (1986): 22–27.

————. *Power/Knowledge: Selected Interviews and Other Writings, 1972–1977.* Translated by C. Gordon. Brighton, U.K.: Harvester, 1980.

————. "Right of Death and Power over Life." In *The Foucault Reader*, edited by P. Rabinow, 258–72. New York: Pantheon, 1984.

Franklin, Sarah. *Biological Relatives: IVF, Stem Cells, and the Future of Kinship.* Durham, NC: Duke University Press, 2013.

————. "Biologization Revisited: Kinship Theory in the Context of the New Biologies." In *Relative Values: Reconfiguring Kinship Studies*, edited by S. Franklin and S. McKinnon, 302–28. Durham, NC: Duke University Press, 2001.

————. "Cyborg Embryo: Our Path to Transbiology." *Theory, Culture and Society* 23, nos. 7–8 (2006): 167–87.

————. "Ethical Biocapital." In *Remaking Life and Death: Toward an Anthropology of the Biosciences*, edited by S. Franklin and M. Lock, 97–128. Santa Fe, NM: School of American Research Press, 2003.

Fuentes, Agustin. "Ethnoprimatology and the Anthropology of the Human-Primate Interface." *Annual Review of Anthropology* 41 (2012): 101–17.

————. "Naturalcultural Encounters in Bali: Monkeys, Temples, Tourists, and Ethnoprimatology." *Cultural Anthropology* 25, no. 4 (2010): 600–624.

Fuentes, Agustin, and Scott Gamerl. "Disproportionate Participation by Age/Sex Classes in Aggressive Interactions between Long-Tailed Macaques (*Macaca fascicularis*) and Human Tourists at Padangtegal Monkey Forest, Bali, Indonesia." *American Journal of Primatology* 66, no. 2 (2005): 197–204.

Fuentes, Agustin, Stephanie Kalchik, Lee Gettler, Anne Kwlatt, and Mckenna Konecki. "Characterizing Human–Macaque Interactions in Singapore." *American Journal of Primatology* 70, no. 9 (2008): 879–83.

Fuentes, Agustin, Eric Shaw, and John Cortes. "Qualitative Assessment of Macaque Tourist Sites in Padangtegal, Bali, Indonesia, and the Upper Rock Nature Reserve, Gibraltar." *International Journal of Primatology* 28, no. 5 (2007): 1143–58.

Fujimura, Joan H. "Authorizing Knowledge in Science and Anthropology." *American Anthropologist* 100, no. 2 (1998): 347–60.

García Canclini, Néstor. *Hybrid Cultures: Strategies for Entering and Leaving Modernity.* Minneapolis: University of Minnesota Press, 2005.

Garmyn, An, Pascale Van Rooij, Frank Pasmans, Tom Hellebuyck, Wim Van Den Broeck, Freddy Haesebrouck, and An Martel. "Waterfowl: Potential Environmental Reservoirs of the Chytrid Fungus *Batrachochytrium dendrobatidis.*" *PLOS ONE* 7, no. 4 (2012): e35038.

Geoghegan, Vincent. "An Anti-humanist Utopia?" In *The Privatization of Hope: Ernst Bloch and the Future of Utopia*, edited by P. Thompson and S. Žižek, 37–60. Durham, NC: Duke University Press, 2013.

Gerhardt, Karin. "Tree Seedling Development in Tropical Dry Abandoned Pasture and Secondary Forest in Costa Rica." *Journal of Vegetation Science* 4 (1993): 95–102.

Gill, Douglas E. "A Naturalist's Guide to the OTS Palo Verde Field Station." Unpublished manuscript, Organization for Tropical Studies, San Jose, Costa Rica, 1988, Biblioteca OET: AD 615.

Gleason, H. A. "The Individualistic Concept of the Plant Association." *Bulletin of the Torrey Botanical Club* 53, no. 1 (1926): 7–26.

Goetsch, C., J. Wigg, A. A. Royo, T. Ristau, and W. P. Carson. "Chronic Over Browsing and Biodiversity Collapse in a Forest Understory in Pennsylvania: Results from a 60 Year-Old Deer Exclusion Plot." *Journal of the Torrey Botanical Society* 138 (2011): 220–24.

Gordon, Linda. *Woman's Body, Woman's Right: A Social History of Birth Control in America.* New York: Penguin, 1977.

Gottschalk, Bob. "Study of the Silver River Monkeys (*Macaca mulatta*)." Unpublished manuscript, 2011.

Gould, Stephen Jay. *The Mismeasure of Man.* New York: Norton, 1981.

———. "Return of Hopeful Monsters." *Natural History* 86, no. 6 (1977): 22–30.

Green, Sherril L. *The Laboratory Xenopus sp.* Boca Raton, FL: CRC Press, 2010.

Gruen, Lori. "Entangled Empathy: An Alternate Approach to Animal Ethics." In *The Politics of Species: Reshaping Our Relationships with Other Animals*, edited by R. Corby and A. Lanjouw, 223–31. Cambridge: Cambridge University Press, 2013.

Gurdon, John B. "*Xenopus* as a Laboratory Animal." In *The Biology of Xenopus*, edited by R. C. Tinsley and H. R. Kobel, 3–8. Oxford: Clarendon, 1996.

Haber, William A., Willow Zuchowski, and Erick Bello. *An Introduction to Cloud Forest Trees: Monteverde, Costa Rica.* Monteverde, Costa Rica: W. Haber, 1996.

Hamaker, Elaine. "Springs Monkeys Preceded Tarzan." *Ocala Star-Banner*, February 1, 1987.

Haraway, Donna. *The Companion Species Manifesto: Dogs, People, and Significant Otherness.* Chicago: Prickly Paradigm Press, 2003.

———. "Manifesto for Cyborgs: Science, Technology, and Socialist Feminism in the 1980s." *Socialist Review* 80 (1985): 65–108.

———. *Modest_Witness@Second_Millennium.FemaleMan_Meets_OncoMouse: Feminism and Technoscience.* New York: Routledge, 1997.

———. *Primate Visions: Gender, Race, and Nature in the World of Modern Science.* London: Verso, 1992.

———. "The Promises of Monsters: A Regenerative Politics for Inappropriate/d Others." In *Cultural Studies*, edited by L. Grossberg, C. Nelson, and P. A. Treichler, 296–337. New York: Routledge, 1992.

———. "Speculative Fabulations for Technoculture's Generations: Taking Care of Unexpected Country." In *The Multispecies Salon*, edited by E. Kirksey, 242–61. Durham, NC: Duke University Press, 2014.

———. *When Species Meet.* Minneapolis: University of Minnesota Press, 2008.

Harding, Susan Friend. "Get Religion." In *The Insecure American: How We Got Here and What We Should Do about It*, edited by H. Gusterson and C. Besteman, 345–61. Berkeley: University of California Press, 2010.

Hardt, Michael, and Antonio Negri. *Multitude: War and Democracy in the Age of Empire*. New York: Penguin, 2004.

Harris, James A. "Do Feedbacks from the Soil Biota Secure Novelty in Ecosystems?" In *Novel Ecosystems: Intervening in the New Ecological World Order*, edited by R. J. Hobbs, E. Higgs, and C. M. Hall, 124–26. Hoboken, NJ: John Wiley, 2013.

Hartigan, John. "Millenials [*sic*] for Obama and the Messy Antic Ends of Race." *Anthropology Now* 2, no. 3 (2009): 1–9.

Harvey, David. *The Condition of Postmodernity: An Enquiry into the Origins of Cultural Change*. Oxford: Blackwell, 1989.

——. "Flexible Accumulation through Urbanization: Reflections on 'Postmodernism' in the American City." *Perspecta* 26 (1990): 251–72.

——. *Spaces of Hope*. Berkeley: University of California Press, 2000.

Hatley, James. "Blood Intimacies and Biodicy: Keeping Faith with Ticks." *Australian Humanities Review* 50 (2011): 63–75.

Hayward, Eva. "FingeryEyes: Impressions of Cup Corals." *Cultural Anthropology* 25, no. 4 (2010): 577–99.

Heatherington, Tracey. "From Ecocide to Genetic Rescue: Can Technoscience Save the Wild?" In *The Anthropology of Extinction: Essays on Culture and Species Death*, edited by G. M. Sodikoff, 39–66. Bloomington: Indiana University Press, 2012.

Helmreich, Stefan. *Alien Ocean: Anthropological Voyages in Microbial Seas*. Berkeley: University of California Press, 2009.

——. "Species of Biocapital." *Science as Culture* 17, no. 4 (2008): 463–78.

——. "Trees and Seas of Information: Alien Kinship and the Biopolitics of Gene Transfer in Marine Biology and Biotechnology." *American Ethnologist* 30, no. 3 (2003): 341–59.

Henson, Pamela M. "Invading Arcadia: Women Scientists in the Field in Latin America, 1900–1950." *The Americas* 58, no. 4 (2002): 577–600.

Herrel, Anthony, and Arie van der Meijden. "An Analysis of the Live Reptile and Amphibian Trade in the USA Compared to the Global Trade in Endangered Species." *Herpetological Journal* 24 (April 2014): 103–10.

Herzog, Hal. *Some We Love, Some We Hate, Some We Eat: Why It's So Hard to Think Straight about Animals*. New York: Harper Perennial, 2010.

Hodgetts, Timothy, and Jamie Lorimer. "Methodologies for Animals' Geographies: Cultures, Communication and Genomics." *Cultural Geographies*, March 3, 2014, 1–11.

Hogben, Lancelot. "*Xenopus* Test for Pregnancy." Correspondence. *British Medical Journal* 2 (July 1939): 38–39.

Holl, Karen D. "Effect of Shrubs on Tree Seedling Establishment in an Abandoned Tropical Pasture." *Journal of Ecology* 90, no. 1 (2002): 179–87.

Hollander, Gail M. *Raising Cane in the 'Glades: The Global Sugar Trade and Transformation of Florida*. Chicago: University of Chicago Press, 2008.

Hölldobler, Bert, and E. O. Wilson. *The Ants*. Cambridge, MA: Harvard University Press, 1990.

Honey, Martha. *Ecotourism and Sustainable Development: Who Owns Paradise?* Washington, DC: Island Press, 2008.

Hora, Riviane R., Chantal Poteaux, Claudie Doums, Dominique Fresneau, and Renée Fénéron. "Egg Cannibalism in a Facultative Polygynous Ant: Conflict for Reproduction or Strategy to Survive?" *Ethology* 113, no. 9 (2007): 909–16.

Hora, R. R., E. Vilela, R. Fénéron, A. Pezon, D. Fresneau, and J. Delabie. "Facultative Polygyny in *Ectatomma tuberculatum* (Formicidae, Ectatomminae)." *Insectes Sociaux* 52, no. 2 (2005): 194–200.

Howard, Walter T. "Vigilante Justice and National Reaction: The 1937 Tallahassee Double Lynching." *Florida Historical Quarterly* 67, no. 1 (1988): 32–51.

Ingold, Tim. *The Perception of the Environment.* New York: Routledge, 2000.

Jaclin, David. "In the (Bleary) Eye of the Tiger: An Anthropological Journey into Jungle Backyards." *Social Science Information Sur Les Sciences Sociales* 52, no. 2 (2013): 257–71.

James, Timothy Y., et al. "Reconstructing the Early Evolution of Fungi Using a Six-Gene Phylogeny." *Nature* 443, no. 7113 (2006): 818–22.

Jameson, Fredric. "Marx's Purloined Letter." In *Ghostly Demarcations: A Symposium on Jacques Derrida's Specters of Marx*, edited by M. Sprinker, 26–67. New York: Verso, 1999.

Janzen, Daniel H. *Costa Rican Natural History.* Chicago: University of Chicago Press, 1983.

———. "Herbivores and the Number of Tree Species in Tropical Forests." *American Naturalist* 104, no. 940 (1970): 501–28.

Janzen, Daniel H., and Paul S. Martin. "Neotropical Anachronisms: The Fruits the Gomphotheres Ate." *Science* 215, no. 4528 (1982): 19–27.

Jeral, J. M., M. D. Breed, and B. E. Hibbard. "Thief Ants Have Reduced Quantities of Cuticular Hydrocarbons in a Ponerine Ant, *Ectatomma ruidum*." *Physiological Entomology* 22 (1997): 207–11.

Journet, Debra. "Ecological Theories as Cultural Narratives: F.E. Clements's and H.A. Gleason's 'Stories' of Community Succession." *Written Communication* 8, no. 4 (1991): 446–72.

Kearns, Carol A., and David W. Inouye. "Pollinators, Flowering Plants, and Conservation Biology." *BioScience* 47, no. 5 (1997): 297–307.

Kennedy, A. C., and L. Z. de Luna. "Rhizosphere." In *Encyclopedia of the Soils in the Environment*, vol. 3, edited by D. Hillel, 399–406. New York: Elsevier, 2005.

Kessler, M. J., and J. K. Hilliard. "Seroprevalence of B Virus (Herpesvirus simiae) Antibodies in a Naturally Formed Group of Rhesus Macaques. *Journal of Medical Primatology* 119 (1990): 155–60.

Khan, Miriam. "Heterotopic Dissonance in the Museum Representation of Pacific Island Cultures." *American Anthropologist* 97, no. 2 (1995): 324–38.

Kirksey, Eben. *Freedom in Entangled Worlds: West Papua and the Architecture of Global Power.* Durham, NC: Duke University Press, 2012.

———, ed. *The Multispecies Salon.* Durham, NC: Duke University Press, 2014.

———. "Species: A Praxiographic Study." *Journal of the Royal Anthropological Institute* 21, no. 4 (2015).

Kirksey, Eben, Brandon Costelloe-Kuehn, and Dorion Sagan. "Life in the Age of Biotechnology." In *The Multispecies Salon*, edited by E. Kirksey, 185–220. Durham, NC: Duke University Press, 2014.

Kirksey, Eben, Dehlia Hannah, Charlie Lotterman, and Lisa Jean Moore. "The Xenopus Pregnancy Test." Unpublished manuscript, 2015.

Kirksey, Eben, and Stefan Helmreich. "The Emergence of Multispecies Ethnography." *Cultural Anthropology* 25, no. 4 (2010): 545–76.

Kirksey, Eben, Craig Schuetze, and Stefan Helmreich. "Introduction: Tactics of Multispecies Ethnography." In *The Multispecies Salon*, edited by E. Kirksey, 1–24. Durham, NC: Duke University Press, 2014.

Kirksey, Eben, Craig Schuetze, Nick Shapiro, Shiho Satsuka, Natasha Myers, Celia Lowe, Jacob Metcalf, Matei Candea, and Stefan Helmereich. "Poaching at the Multispecies Salon." *Kroeber Anthropological Society Papers* 99/100 (2011): 129–53.

Kirksey, Eben, Nicholas Shapiro, and Maria Brodine. "Hope in Blasted Landscapes." In *The Multispecies Salon*, edited by E. Kirksey, 29–63. Durham, NC: Duke University Press, 2014.

Knight, John. *Animals in Person: Cultural Perspectives on Human-Animal Intimacy.* Oxford: Berg, 2005.

Kockelman, Paul. "Enemies, Parasites, and Noise: How to Take Up Residence in a System without Becoming a Term in It." *Journal of Linguistic Anthropology* 20, no. 2 (2010): 406–21.

———. "A Mayan Ontology of Poultry: Selfhood, Affect, Animals, and Ethnography." *Language in Society* 40 (2011): 1–28.

Kohler, Robert E. *All Creatures: Naturalists, Collectors, and Biodiversity, 1850–1950.* Princeton, NJ: Princeton University Press, 2006.

———. *Lords of the Fly: Drosophila Genetics and the Experimental Life.* Chicago: University of Chicago Press, 1994.

Kohn, Eduardo. *How Forests Think: Toward an Anthropology beyond the Human.* Berkeley: University of California Press, 2013.

Koptur, Suzanne. "Breeding Systems of Monteverde *Inga*." In *Monteverde: Ecology and Conservation of a Tropical Cloud Forest*, edited by N. M. Nadkarni and N. T. Wheelwright, 85–87. New York: Oxford University Press, 2000.

———. "Interactions among *Inga*, Herbivores, Ants, and Insect Visitors to Foliar Nectaries." In *Monteverde: Ecology and Conservation of a Tropical Cloud Forest*, edited by N. M. Nadkarni and N. T. Wheelwright, 277–78. New York: Oxford University Press, 2000.

Kosek, Jake. "Ecologies of Empire: On the New Uses of the Honeybee." *Cultural Anthropology* 25, no. 4 (2010): 650–78.

Kowal, Emma, and Joanna Radin. "Indigenous Biospecimen Collections and the Cryopolitics of Frozen Life." *Journal of Sociology* 51, no. 1 (2015).

Lachaud, Jean-Paul, Alex Cadena, Bertrand Schatz, Gabriela Perez-Lachaud, and

Guillermo Ibarra-Nunez. "Queen Dimorphism and Reproductive Capacity in the Ponerine Ant, *Ectatomma ruidum* Roger." *Oecologia* 120 (1999): 515–23.

Lamb, David, John Parrotta, Rod Keenan, and Nigel Tucker. "Rejoining Habitat Remnants: Restoring Degraded Rainforest Lands." In *Tropical Forest Remnants: Ecology, Management, and Conservation of Fragmented Communities*, edited by W. F. Laurance and R. O. Bierregaard, 366–85. Chicago: University of Chicago Press, 1997.

Latour, Bruno. *Politics of Nature*. Cambridge, MA: Harvard University Press, 2004.

———. *Reassembling the Social: An Introduction to Actor-Network-Theory*. Oxford: Oxford University Press, 2005.

———. *Science in Action*. Cambridge, MA: Harvard University Press, 1987.

Lawrence, Elizabeth Atwood. *Rodeo: An Anthropologist Looks at the Wild and the Tame*. Chicago: University of Chicago Press, 1984.

Layne, James N. "Nonindigenous Mammals." In *Strangers in Paradise: Impact and Management of Nonindigenous Species in Florida*, edited by D. Simberloff, D. C. Schmitz, and T. C. Brown, 157–86. Washington, DC: Island Press, 1997.

Leavitt, Sarah Abigail. "'A Private Little Revolution': The Home Pregnancy Test in American Culture." *Bulletin of the History of Medicine* 80, no. 2 (2006): 317–45.

Leigh, Egbert Giles. *Tropical Forest Ecology: A View from Barro Colorado Island*. New York: Oxford University Press, 1999.

Lestel, Dominique, F. Brunois, and F. Gaunet. "Etho-ethnology and Ethno-ethology." *Social Science Information Sur Les Sciences Sociales* 45, no. 2 (2006): 155–77.

Leston, D. "The Ant Mosaic, Tropical Tree Crops, and the Limiting of Pests and Diseases." *Pest Articles and New Summaries* 19 (1973): 311–41.

Lévinas, Emmanuel. *Otherwise Than Being, or, Beyond Essence*. Pittsburgh: Duquesne University Press, 1998.

Lévi-Strauss, Claude. *The Savage Mind*. Chicago: University of Chicago Press, 1966.

———. *Totemism*. Boston: Beacon, 1963.

Lindsay-Poland, John. *Emperors in the Jungle: The Hidden History of the U.S. in Panama*. Durham, NC: Duke University Press, 2003.

Lowe, Celia. "Viral Clouds: Becoming H5N1 in Indonesia." *Cultural Anthropology* 25, no. 4 (2010): 625–49.

———. *Wild Profusion: Biodiversity Conservation in an Indonesian Archipelago*. Princeton, NJ: Princeton University Press, 2006.

Lyons, Kristina Marie. "Soil Science, Development, and the 'Elusive Nature' of Colombia's Amazonian Plains." *Journal of Latin American and Caribbean Anthropology* 19, no. 2 (2014): 212–36.

Macfadyen, James. *The Flora of Jamaica*. London: Longman, Orme, Brown, Green, and Longman, 1837.

MacNeill, B. H. "Parasexuality in Plant Pathogenic Fungi." *Annual Review of Phytopathology* 7 (1969): 147–69.

Maestripieri, Dario. *Macachiavellian Intelligence: How Rhesus Macaques and Humans Have Conquered the World*. Chicago: University of Chicago Press, 2007.

Marcus, George E. *Para-sites: A Casebook against Cynical Reason*. Chicago: University of Chicago Press, 2000.

Marder, Michael. *Plant Thinking: A Philosophy of Vegetal Life*. New York: Columbia University Press, 2013.

Margulis, Lynn, and Dorion Sagan. *Acquiring Genomes: A Theory of the Origins of Species*. New York: Basic Books, 2002.

Marks, Jonathan. "'We're Going to Tell These People Who They Really Are': Science and Relatedness." In *Relative Values: Reconfiguring Kinship Studies*, edited by S. Franklin and S. McKinnon, 355–83. Durham, NC: Duke University Press, 2001.

Marx, Karl. *Capital*, vol. 1. London: Penguin Classics, 1976.

Mascaro, Joseph. "Perspective: From Rivets to Rivers." In *Novel Ecosystems: Intervening in the New Ecological World Order*, edited by R. J. Hobbs, E. Higgs, and C. M. Hall, 155–56. Hoboken, NJ: John Wiley, 2013.

Mascaro, Joseph, James A. Harris, Lori Lach, Allen Thompson, Michael P. Perring, David M. Richardson, and Erle C. Ellis. "Origins of the Novel Ecosystems Concept." In *Novel Ecosystems: Intervening in the New Ecological World Order*, edited by R. J. Hobbs, E. Higgs, and C. M. Hall, 45–57. Hoboken, NJ: John Wiley, 2013.

Masco, Joseph. "Mutant Ecologies: Radioactive Life in Post–Cold War New Mexico." *Cultural Anthropology* 19, no. 4 (2004): 517–50.

Massumi, Brian. *Parables for the Virtual: Movement, Affect, Sensation*. Durham, NC: Duke University Press, 2002.

Matsutake Worlds Research Group. "A New Form of Collaboration in Cultural Anthropology: Matsutake Worlds." *American Ethnologist* 36, no. 2 (2009): 380–403.

———. "Thoughts for a World of Poaching." *Cultural Anthropology*, October 10, 2010. http://www.culanth.org/fieldsights/276-thoughts-for-a-world-of-poaching.

Maxwell, Greg. *The More Complete Chondro: A Comprehensive Guide to the Care and Breeding of Green Tree Pythons*. Rodeo, NM: ECO Herpetological Publishing and Distribution, 2005.

McCoy, Michael B. "Seasonal, Freshwater Marshes in the Tropics: A Case in Which Cattle Grazing Is Not Detrimental." In *Principles of Conservation Biology*, edited by G. K. Meffe and C. R. Carroll, 352–53. Sunderland, MA: Sinauer, 1994.

McCoy, Michael B., and J. M. Rodriguez Ramirez. "Cattail (*Typha domingensis*) Eradication Methods in the Restoration of a Tropical, Seasonal, Freshwater Marsh." In *Global Wetlands: Old World and New*, edited by W. J. Mitsch, 469–82. Amsterdam: Elsevier Service, 1994.

McMahon, Taegan A., Laura A. Brannelly, Matthew W. H. Chatfield, Pieter T. J. Johnson, Maxwell B. Joseph, Valerie J. McKenzie, Corinne L. Richards-Zawacki, Matthew D. Venesky, and Jason R. Rohr. "Chytrid Fungus *Batrachochytrium dendrobatidis* Has Nonamphibian Hosts and Releases Chemicals That Cause Pathology in the Absence of Infection." *Proceedings of the National Academy of Sciences* 110, no. 1 (2013): 210–15.

Mehlman, P. T., J. D. Higley, B. J. Fernald, F. R. Sallee, S. J. Suomi, and M. Linnoila. "CSF 5-HIAA, Testosterone, and Sociosexual Behaviors in Free-Ranging Male Rhesus Macaques in the Mating Season." *Psychiatry Research* 72, no. 2 (1997): 89–102.

Mello, Anna J., Patricia A. Townsend, and Katie Filardo. "Reforestation and the Restoration at the Cloud Forest School in Monteverde, Costa Rica: Learning by Doing." *Ecological Restoration* 28, no. 2 (2010): 148–50.

Melson, W. "Prehistoric Eruptions of Arenal Volcano, Costa Rica." *Vinculos* 10 (1984): 35–59.

Millikan, Max. "Pareto's Sociology." *Econometrica* 4, no. 4 (1936): 324–37.

Mitchell, Timothy. *Rule of Experts: Egypt, Techno-politics, Modernity*. Berkeley: University of California Press, 2002.

Mol, Annemarie. *The Body Multiple: Ontology in Medical Practice*. Durham, NC: Duke University Press, 2002.

Montague, Clay L., Sheila V. Colwell, H. Franklin Percival, and Johan F. Gottgens. "Issues and Options Related to Management of Silver Springs Rhesus Macaques." *National Biological Survey Technical Report* 49 (1994): 34.

Mooallem, Jon. "What's a Monkey to Do in Tampa?" *New York Times Magazine*, August 22, 2012.

Moore, Lisa Jean, and Mary Kosut. *Buzz: Urban Beekeeping and the Power of the Bee*. New York: New York University Press, 2013.

Morel, Laura C. "Trappers Use Fruit, Sandwiches to Entice Mystery Monkey of Tampa Bay." *Tampa Bay Times*, October 10, 2012.

Moss, Angela S., Nikla S. Reddy, Ida M. Dortaj, and Michael J. San Francisco. "Chemotaxis of the Amphibian Pathogen *Batrachochytrium dendrobatidis* and Its Response to a Variety of Attractants." *Mycologia* 100, no. 1 (2008): 1–5.

Mullin, Molly. "Mirrors and Windows: Sociocultural Studies of Human-Animal Relationships." *Annual Review of Anthropology* 28 (1999): 201–24.

Myers, Natasha. "Performing the Protein Fold." In *Simulations and Its Discontents*, edited by S. Turkle, 171–201. Cambridge, MA: MIT Press, 2009.

Nadkarni, Nalini, and Nathaniel T. Wheelwright. *Monteverde: Ecology and Conservation of a Tropical Cloud Forest*. New York: Oxford University Press, 2000.

Netz, Reviel. *The Cutting Edge: An Environmental History of Modernity*. Middletown, CT: Wesleyan University Press, 2004.

Ogden, Laura A. *Swamplife: People, Gators, and Mangroves Entangled in the Everglades*. Minneapolis: University of Minnesota Press, 2011.

Oliver, James H. "Biology and Systematics of Ticks (Acari: Ixodida)." *Annual Review of Ecology and Systematics* 20 (1989): 397–430.

Olszynko-Gryn, Jesse. "When Pregnancy Tests Were Toads: The Xenopus Test in the Early NHS." *Wellcome History* 51 (2013): 2–3.

Ortiz, Fernando. *Cuban Counterpoint: Tobacco and Sugar*. 1940. Reprint, Durham, NC: Duke University Press, 1995.

Palaversich, Diana. *De Macondo a McOndo*. Barcelona: Plaza y Valdes, 2005.

Parreñas, Rheana Salazar. "Producing Affect: Transnational Volunteerism in a

Malaysian Orangutan Rehabilitation Center." *American Ethnologist* 39, no. 4 (2012): 673–87.

Paxson, Heather. "Microbiopolitics." In *The Multispecies Salon*, edited by E. Kirksey, 115–21. Durham, NC: Duke University Press, 2014.

Perfecto, I., and J. H. Vandermeer. "Cleptobiosis in the Ant *Ectatomma ruidum* in Nicaragua." *Insectes Sociaux* 40 (1993): 295–99.

Peterson, C. J., and B. L. Haines. "Early Successional Patterns and Potential Facilitation of Woody Plant Colonization by Rotting Logs in Premontane Costa Rican Pastures." *Restoration Ecology* 8, no. 4 (2000): 361–69.

Peterson, Roger Tory. *Peterson Field Guide to Birds of Eastern and Central North America*. Boston: Houghton Mifflin Harcourt, 2010.

Plumwood, Val. *Feminism and the Mastery of Nature*. London: Routledge, 1993.

Popielarz, P. A., and Z. P. Neal. "The Niche as a Theoretical Tool." *Annual Review of Sociology* 33 (2007): 65–84.

Pratt, Mary Louise. *Imperial Eyes: Travel Writing and Transculturation*. London: Routledge, 1992.

Pratt, Stephen C. "Recruitment and Other Communication Behavior in the Ponerine Ant *Ectatomma ruidum*." *Ethology* 81 (1989): 313–31.

Puig de la Bellacasa, Maria. "'Nothing Comes without Its World': Thinking with Care." *Sociological Review* 60, no. 2 (2012): 197–216.

Qu, W. Y., Y. Z. Zhang, D. Manry, and C. H. Southwick. "Rhesus Monkeys (*Macaca mulatta*) in the Taihang Mountains, Jiyuan County, Henan, China." *International Journal of Primatology* 14, no. 4 (1993): 607–21.

Rabinow, Paul, and Nikolas Rose. "Biopower Today." *BioSocieties* 1 (2006): 195–217.

Raffles, Hugh. *The Illustrated Insectopedia: Insect Love from A–Z*. New York: Pantheon/Vintage, 2010.

Reznikova, Zhanna I. *Animal Intelligence: From Individual to Social Cognition*. Cambridge: Cambridge University Press, 2007.

Richard, T. L. "Compost." In *Encyclopedia of the Soils in the Environment*, vol. 1, edited by D. Hillel, 294–300. New York: Elsevier, 2005.

Riley, Erin P. "Contemporary Primatology in Anthropology: Beyond the Epistemological Abyss." *American Anthropologist* 115, no. 3 (2013): 411–22.

Ritvo, Harriet. *The Animal Estate: The English and Other Creatures in the Victorian Age*. Cambridge, MA: Harvard University Press, 1987.

Rizo, Federico. "Monitoreo de los Arrozales del Proyecto Tamarindo: Estudio de los Agroquimicos y Macroinvertebrados Bentonicos en Relacion al Parque Nacional Palo Verde." Paper presented at Restauracion y Conservacion de Ecosistemas en la Cuenca baja del Rio Tempisque: Hacia una Perspectiva de Manejo Integrado, Bagaces, Guanacaste, November 6–7, 2003.

Ronan, Charles E. "Observations on the Word Gringo." *Arizona and the West* 6, no. 1 (1964): 23–29.

Rose, Deborah Bird. *Reports from a Wild Country: Ethics for Decolonisation*. Sydney: UNSW Press, 2004.

————. "What If the Angel of History Were a Dog?" *Cultural Studies Review* 12, no. 1 (2006): 67–74.

————. *Wild Dog Dreaming: Love and Extinction.* Charlottesville: University of Virginia Press, 2011.

⚡Rose, Deborah Bird, and Thom van Dooren. "Unloved Others: Death of the Disregarded in the Time of Extinctions." *Australian Humanities Review* 50 (2011).

Rosenblum, E. B., et al. "Complex History of the Amphibian-Killing Chytrid Fungus Revealed with Genome Resequencing Data." *Proceedings of the National Academy of Sciences* 110, no. 23 (2013): 9385–90.

Rotman, Brian. *Becoming Beside Ourselves: The Alphabet, Ghosts, and Distributed Human Being.* Durham, NC: Duke University Press, 2008.

Rutherford, Danilyn. *Laughing at Leviathan: Sovereignty and Audience in West Papua.* Chicago: University of Chicago Press, 2012.

Sagan, Dorion. "Umwelt after Uexküll." In *A Foray into the Worlds of Animals and Humans,* with *A Theory of Meaning,* by Jakob von Uexküll. Minneapolis: University of Minnesota Press, 2010.

Sanchez, Julio, Juan M. Rodriguez, and Carlos Salas. *Distribución, ciclos reproductivos y aspectos ecológicos de aves acuáticas.* San Jose, Costa Rica: EUNED, 1985.

Santer, Eric L. *On Creaturely Life.* Chicago: University of Chicago Press, 2006.

Sawyer, Charles. *Our Sport: Market Hunting.* Los Banos: Loose Change, 2005.

Schatz, B., G. Beugnon, and J. P. Lachaud. "Time-Place Learning by an Invertebrate, the Ant *Ectatomma-Ruidum* Roger." *Animal Behaviour* 48, no. 1 (1994): 236–38.

Schatz, B., J. P. Lachaud, G. Beugnon, and A. Dejean. "Prey Density and Polyethism within Hunting Workers in the Neotropical Ponerine Ant *Ectatomma ruidum* (Hymenoptera, Formicidae)." *Sociobiology* 34, no. 3 (1999): 605–17.

Schloegel, L. M., L. F. Toledo, J. E. Longcore, S. E. Greenspan, C. A. Vieira, M. Lee, S. Zhao, C. Wangen, C. M. Ferreira, M. Hipolito, A. J. Davies, C. A. Cuomo, P. Daszak, and T. Y. James. "Novel, Panzootic and Hybrid Genotypes of Amphibian Chytridiomycosis Associated with the Bullfrog Trade." *Molecular Ecology* 21 (2012): 5162–77.

Schneider, David Murray. *A Critique of the Study of Kinship.* Ann Arbor: University of Michigan Press, 1984.

Schuck-Paim, Cynthia, Lorena Pompilio, and Alex Kacelnik. "State-Dependent Decisions Cause Apparent Violations of Rationality in Animal Choice." *PLOS Biology* 2, no. 12 (2004): e402.

Serres, Michel. *The Parasite.* Minneapolis: University of Minnesota Press, 2007.

Shapin, Steven, and Simon Schaffer. *Leviathan and the Air-Pump: Hobbes, Boyle, and the Experimental Life.* Princeton, NJ: Princeton University Press, 1985.

Sheets, Payson, John Hoopes, William Melson, Brian McKee, Tom Sever, Marilynn Mueller, Mark Chenault, and John Bradley. "Prehistory and Volcanism in the Arenal Area, Costa Rica." *Journal of Field Archaeology* 18 (1991): 445–65.

Shik, J. Z., and M. Kaspari. "Lifespan in Male Ants Linked to Mating Syndrome." *Insectes Sociaux* 56 (2009): 131–34.

Shir-Vertesh, Dafna. "'Flexible Personhood': Loving Animals as Family Members in Israel." *American Anthropologist* 114, no. 3 (2012): 420–32.

Shukin, Nicole. *Animal Capital: Rendering Life in Biopolitical Times*. Minneapolis: University of Minnesota Press, 2009.

Silk, J. B. "Male Bonnet Macaques Use Information about Third-Party Rank Relationships to Recruit Allies." *Animal Behaviour* 58 (1999): 45–51.

Simmons, D. Rabern, and Joyce E. Longcore. "*Thoreauomyces* gen. nov., *Fimicolochytrium* gen. nov. and Additional Species in *Geranomyces*." *Mycologia* 104, no. 5 (2012): 1229–43.

Siskind, Janet. "Kinship and Mode of Production." *American Anthropologist* 80, no. 4 (1978): 860–72.

Sleigh, Charlotte. *Six Legs Better: A Cultural History of Myrmecology*. Baltimore: Johns Hopkins University Press, 2007.

Sloterdijk, Peter. "Atmospheric Politics." In *Making Things Public: Atmospheres of Democracy*, edited by B. Latour and P. Weibel, 944–52. Cambridge, MA: MIT Press, 2005.

———. *Bubbles: Spheres*. Vol. 1, *Microspherology*. Cambridge, MA: MIT Press, 2011.

Sodikoff, Genese Marie. *The Anthropology of Extinction: Essays on Culture and Species Death*. Bloomington: Indiana University Press, 2012.

Srinivasan, Krithika. "Caring for the Collective: Biopower and Agential Subjectification in Wildlife Conservation." *Environment and Planning D: Society and Space* 32, no. 3 (2014): 501–17.

Standish, Rachel J., Allen Thompson, Eric Higgs, and Stephen D. Murphy. "Concerns about Novel Ecosystems." In *Novel Ecosystems: Intervening in the New Ecological World Order*, edited by R. J. Hobbs, E. Higgs, and C. M. Hall, 296–309. Hoboken, NJ: John Wiley, 2013.

Star, Susan Leigh, and James R. Griesemer. "Institutional Ecology, 'Translation,' and Boundary Objects: Amateurs and Professionals in Berkeley's Museum of Vertebrate Zoology, 1907–1939." *Social Studies of Science* 19 (1989): 387–420.

Stengers, Isabelle. "The Cosmopolitical Proposal." In *Making Things Public: Atmospheres of Democracy*, edited by B. Latour and P. Weibel, 994–1003. Cambridge, MA: MIT Press, 2005.

———. *Cosmopolitics I*. Minneapolis: University of Minnesota Press, 2010.

———. *Cosmopolitics II*. Minneapolis: University of Minnesota Press, 2011.

Stepan, Nancy Leys. *Picturing Tropical Nature*. Ithaca, NY: Cornell University Press, 2001.

Stoetzer, Bettina. *At the Forest Edges of the City: An Ethnography of Racial Geographies and National Belonging in Berlin*. Berkeley: University of California Press, 2011.

Strain, Ellen. "Stereoscopic Visions: Touring the Panama Canal." *Visual Anthropology Review* 12, no. 2 (1996/1997): 44–58.

Strauss, Anselm. "A Social World Perspective." *Studies in Symbolic Interaction* 1 (1978): 119–28.

Stuart, Simon N., Janice S. Chanson, Neil A. Cox, Bruce E. Young, Ana S. L. Rodrigues, Debra L. Fischman, and Robert W. Waller. "Status and Trends of Amphibian Declines and Extinctions Worldwide." *Science* 306, no. 5702 (2004): 1783–86.

Sundberg, Juanita. "Conservation Encounters: Transculturation in the 'Contact Zones' of Empire." *Cultural Geography* 13, no. 2 (2006): 239–65.

Sunder Rajan, Kaushik. *Biocapital: The Constitution of Postgenomic Life*. Durham, NC: Duke University Press, 2006.

Swei, A., et al. "Is Chytridiomycosis an Emerging Infectious Disease in Asia?" *PLOS ONE* 6, no. 8 (2011): e23179.

Tansley, A. G. "The Use and Abuse of Vegetational Concepts and Terms." *Ecology* 16 (1935): 284–307.

Taussig, Michael T. *Shamanism, Colonialism, and the Wild Man: A Study in Terror and Healing*. Chicago: University of Chicago Press, 1986.

ten Bos, Rene. "Towards an Amphibious Anthropology: Water and Peter Sloterdijk." *Environment and Planning D: Society and Space* 27 (2009): 1–11.

Thacker, Eugene. 2004. "Networks, Swarms, Multitudes." *CTheory* no. a142a, May 18, 2004. http://www.ctheory.net/articles.aspx?id=422.

Thierry, Bernard, Singh Mewa, and Werner Kaumanns. *Macaque Societies: A Model for the Study of Social Organization*. Cambridge: Cambridge University Press, 2004.

Thompson, Charis. *Making Parents: The Ontological Choreography of Reproductive Technologies*. Cambridge, MA: MIT Press, 2005.

Thompson, Peter. "The Privatization of Hope and the Crisis of Negation." In *The Privatization of Hope: Ernst Bloch and the Future of Utopia*, edited by P. Thompson and S. Žižek, 1–20. Durham, NC: Duke University Press, 2013.

Tilley, Helen. *Africa as a Living Laboratory: Empire, Development, and the Problem of Scientific Knowledge, 1870–1950*. Chicago: University of Chicago Press, 2011.

Tinsley, R. C. "Parasites of *Xenopus*." In *The Biology of Xenopus*, edited by R. C. Tinsley and H. R. Kobel, 233–59. Oxford: Clarendon, 1996.

Tinsley, R. C., Leslie H. Chappell, and H. R. Kobel. "Geographical Distribution and Ecology." In *The Biology of Xenopus*, edited by R. C. Tinsley and H. R. Kobel, 35–59. Oxford: Clarendon, 1996.

Tinsley, R. C., and M. J. McCoid. "Feral Populations of *Xenopus* outside Africa." In *The Biology of Xenopus*, edited by R. C. Tinsley and H. R. Kobel, 143–76. Oxford: Clarendon, 1996.

Townsend, Patricia A. "Conservation and Restoration for a Changing Climate in Monteverde, Costa Rica." PhD diss., University of Washington, Seattle, 2011.

Trama, Florencia Andrea. "Manejo Activo y Restauracion del Humedal Palo Verde: Cambios en las Coberturas de Vegetacion y Respuesta de las Aves Acuaticas." Master's thesis, Universidad Nacional, Heredia, Costa Rica, 2005.

Tsing, Anna. "Blasted Landscapes (and the Gentle Art of Mushroom Picking)." In *The Multispecies Salon*, edited by E. Kirksey, 87–109. Durham, NC: Duke University Press, 2014.

————. "Contaminated Diversity in 'Slow Disturbance': Potential Collaborators for a Livable Earth." *RCC Perspectives* 9 (2012): 95–97.

————. "Indigenous Voice." In *Indigenous Experience Today*, edited by M. de la Cadena and O. Starn. New York: Berg, 2007.

————. "Worlding the Matsutake Diaspora: Or, Can Actor-Network Theory Experiment with Holism." In *Experiments in Holism: Theory and Practice in Contemporary Anthropology*, edited by T. Otto and N. Bubandt. Hoboken, NJ: Wiley-Blackwell, 2010.

Tsing, Anna Lowenhaupt, and Elizabeth Pollman. "Global Futures: The Game." In *Histories of the Future*, edited by D. Rosenberg and S. F. Harding, 107–22. Durham, NC: Duke University Press, 2005.

U.S. Department of the Interior, Fish and Wildlife Service, U.S. Department of Commerce, and U.S. Census Bureau. *National Survey of Fishing, Hunting, and Wildlife-Associated Recreation*. Washington, DC: U.S. Fish and Wildlife Service, 2006.

van den Akker, J. J. H., and B. Soane. "Compaction." In *Encyclopedia of the Soils in the Environment*, vol. 1, edited by D. Hillel, 285–93. New York: Elsevier, 2005.

van Dooren, Thom. *Flight Ways: Life at the Edge of Extinction*. New York: Columbia University Press, 2014.

————. "Invasive Species in Penguin Worlds: An Ethical Taxonomy of Killing for Conservation." *Conservation and Society* 9, no. 4 (2011): 286–98.

————. "Vultures and Their People in India: Equity and Entanglement in a Time of Extinction." *Manoa* 22, no. 2 (2010): 130–46.

Ventocilla, Jorge, and Kurt Dillon. *Gamboa: A Guide to Its Natural and Cultural Heritage*. Panama: Smithsonian Tropical Research Institute, 2010.

Villalobos-Brenes, F., L. Castillo, and J. A. Morales. "Efectos Toxicos y Alteracion Endocrina en la Ictiofauna del Area Bagatzi-Poza Verde." Paper presented at Restauracion y Conservacion de Ecosistemas en la Cuenca baja del Rio Tempisque: Hacia una Perspectiva de Manejo Integrado, Bagaces, Guanacaste, November 6–7, 2003.

Vivanco, Luis. *Green Encounters: Shaping and Contesting Environmentalism in Rural Costa Rica*. New York: Berghahn, 2006.

————. "Spectacular Quetzals, Ecotourism, and Environmental Futures in Monte Verde, Costa Rica." *Ethnology* 40, no. 2 (2001): 79–92.

————. "The Work of Environmentalism in an Age of Televisual Adventures." *Cultural Dynamics* 16, no. 1 (2004): 5–27.

Vivanco, Luis Antonio, and Robert J. Gordon. *Tarzan Was an Eco-tourist: And Other Tales in the Anthropology of Adventure*. New York: Berghahn, 2006.

von Uexküll, Jakob. *A Foray into the Worlds of Animals and Humans*, with *A Theory of Meaning*. Minneapolis: University of Minnesota Press, 2010.

————. "A Stroll through the Worlds of Animals and Men: A Picture Book of Invisible Worlds." *Semiotica* 89, no. 4 (1992 [1934]): 319–91.

Voyles, J., S. Young, L. Berger, C. Campbell, W. F. Voyles, A. Dinudom, D. Cook, R. Webb, R. A. Alford, L. F. Skerratt, and R. Speare. "Pathogenesis of Chytridiomycosis, a Cause of Catastrophic Amphibian Declines." *Science* 326, no. 5952 (2009): 582–85.

Vredenburg, Vance T., Stephen A. Felt, Erica C. Morgan, Samuel V. G. McNally, Sabrina Wilson, and Sherril L. Green. "Prevalence of *Batrachochytrium dendrobatidis* in *Xenopus* Collected in Africa (1871–2000) and in California (2001–2010)." *PLOS ONE* 8, no. 5 (2013): e63791.

Wald, Priscilla. *Contagious: Cultures, Carriers, and the Outbreak Narrative*. Durham, NC: Duke University Press, 2008.

Wallerstein, Immanuel Maurice. *The Decline of American Power: The U.S. in a Chaotic World*. New York: New Press, 2003.

Warkentin, Traci. "Interspecies Etiquette: An Ethics of Paying Attention to Animals." *Ethics and the Environment* 15, no. 1 (2010): 101–21.

Weber, Neal A. "Two Common Ponerine Ants of Possible Economic Significance, *Ectatomma tuberculatum* (Olivier) and *E. ruidum* (Roger)." *Proceedings of the Entomological Society of Washington* 48, no. 1 (1946): 1–16.

Weisser, W. W., and E. Siemann. *Insects and Ecosystem Function*. Berlin: Springer, 2007.

Weldon, Ché. "Chytridiomycosis, an Emerging Infectious Disease of Amphibians in South Africa." PhD diss., North-West University South Africa, 2005.

Weldon, Ché, Louis H. du Preez, Alex D. Hyatt, Reinhold Muller, and Rick Speare. "Origin of the Amphibian Chytrid Fungus." *Emerging Infectious Diseases* 10, no. 12 (2004): 2100–2105.

Wesley, David E. "Socio-duckonomics." In *Valuing Wildlife: Economic and Social Perspectives*, edited by J. D. Decker and G. R. Goff, 136–42. Boulder, CO: Westview, 1987.

Weston, Kath. *Families We Choose: Lesbians, Gays, Kinship*. New York: Columbia University Press, 1991.

Whatmore, Sarah, and L. Thorne. "Wild(er)ness: Reconfiguring the Geographies of Wildlife." *Transactions of the Institute of British Geographers* 23, no. 4 (1998): 435–54.

White, Hayden V. *Figural Realism: Studies in the Mimesis Effect*. Baltimore, MD: Johns Hopkins University Press, 1999.

Williams, Raymond. "Utopia and Science Fiction." *Science Fiction Studies* 5, no. 3 (1978): 203–14.

Williamson, Mark, and Alastair Fitter. "The Varying Success of Invaders." *Ecology* 77, no. 6 (1996): 1661–66.

Wilson, E. O. *On Human Nature*. Cambridge, MA: Harvard University Press, 1988.

Wolfe, Cary. "Introduction." In *The Parasite*, by M. Serres. Minneapolis: University of Minnesota Press, 2007.

———. *What Is Posthumanism?* Minneapolis: University of Minnesota Press, 2010.

Wolfe, Linda D., and Elizabeth H. Peters. "History of the Freeranging Rhesus Monkeys (*Macca mulatta*) of Silver Springs." *Florida Scientist* 50, no. 4 (1987): 234–45.

Wood, David. "On Being Haunted by the Future." *Research in Phenomenology* 26 (2006): 274–98.

Youatt, Rafi. "Counting Species: Biopower and the Global Biodiversity Census."
  Environmental Values 17 (2008): 393–417.
Yung, Laurie, Steve Schwarze, Wylie Carr, F. Stuart Chapin, and Emma Marris.
  "Engaging the Public in Novel Ecosystems." In *Novel Ecosystems: Intervening
  in the New Ecological World Order*, edited by R. J. Hobbs, E. Higgs, and C. M.
  Hall, 247–56. Hoboken, NJ: John Wiley, 2013.
Zurr, Ionat. "Complicating Notions of Life: Semi Living Entities." In *Biomediale:
  Contemporary Society and Genomic Culture*, edited by D. Bulatov, 402–11. Ka-
  liningrad, Russia: National Center for Contemporary Arts, 2004.

INDEX

actor-network theory, 25, 166. *See also* agents; *interessement*

Adams, Wendy: Florida kitsch, 253n85

affect: contagious, 5, 56, 121–23; pack animals and, 133; wildness and, 106–7

affect aliens: defined, 240n41; endangered animals as, 59–60; pets as, 146

affective attachments, 5

Africa: disease outbreak and, 90–91, 99; "out of Africa," 247n13

African Americans, and segregation, 111–13

African clawed frog. See *Xenopus laevis*

Agamben, Giorgio: *umwelt*, 233n12

agents: actor-network theory and, 25; capitalism and, 168, 180; cosmopolitical assembly and, 18, 24, 33; emergent ecosystems and, 3, 190, 217; *interessement* and, 25, 27, 31, 196–97; intra-action and, 78, 84, 95, 246n34; natural vs. cultural, 130

aggression, 15, 27, 31; dominance hierarchies and, 119

agora, 166

agriculture: government sponsored, 178–79; harmless parasites and, 17–18; labor and, 13; nature reserves and, 10, 17, 171; organic farming, 190; pests and, 13, 181–85; subsistence and imperialism, 10, 50–51; swidden, 12; tractors used in conservation, 174;

volcanic eruptions and, 214–15. *See also* campesinos

Ahmed, Sara: affect aliens, 59; happy objects, 56; sticky objects, 40, 56

alliance: cosmopolitical, 18; Deleuze and, 106; fleeting, 123–24; reforestation and, 202. *See also* collaboration

alterity: fetishism and, 136; wildness and, 67–69, 107

altruism, 31. *See also* trophallaxis

amateurs, 5. *See also* Do-It-Yourself (DIY)

Amphibian Ark, 44, 57–59, 62; carrying capacity of, 66; visions of salvation, 70, 154; Year of the Frog capital campaign, 57–58

amphibians: Amphibian Avenger blog, 52; El Valle Amphibian Conservation Center (EVACC), 52–57; endangered, 44, 58, 157; Panama Amphibian Rescue and Conservation Project, 46; redback salamanders (*Plethodon cinereus*), 69–70. *See also* frogs; ontological amphibians

animal rhizomes, 169, 186

animals: agoutis, 36–37, 195, 205, 209, 211; Baird's tapirs (*Tapirus bairdii*), 38; becoming animal, 229n25; chimpanzees, 34, 111–12; commodified, 137, 146; ethology and, 6; experimentation, 94–97; familiars of capital, 168–69; feeding, 117–20, 124;

emergent diseases, 1; frog pathogens, 44, 56, 78–79, 84

emergent ecologies, 3, 4–5, 12, 31, 190; building, 220; care and, 7, 159–61, 173; compost and, 200–202; endangered species and, 68–69, 156–57; exceeding human visions, 195; human households as, 135, 153–58; introduced species and, 104, 130–31, 143, 151; nomadic animals invade, 17–18, 97; parasites shape, 100, 184–86; reforestation and, 209; volcanic eruptions and, 214–15; wildness and, 136–37. *See also* ecosystem

emergent plants, 203–4, 210–11

empathy (Gruen), 34

endangered species, 5–6, 46; banded horned tree frog (*Hemiphractus fasciatus*), 55; blue-sided tree frog (*Agalychnis annae*), 158–61, 258n73; confusion about legislation, 160; Department of Defense and, 238n9; flourishing in new circumstances, 143, 156–57, 188; frogs, 44–47, 55, 239n30; golden frogs (*Atelopus zeteki*), 51, 57–66; Golfodulcean poison frog (*Phyllobates vittatus*), 156–57, 258n65; habitat loss and, 55, 68; harlequin frogs (*Atelopus* species), 46, 240n35; killed by conservationists, 60, 173; lemur leaf frogs (*Agalychnis lemur*), 53; plants, 213; sun parakeets (golden conures or *Aratinga solstitialis*), 141, 143; trade in, 65–66, 134–36. *See also* extinction

Endangered Species Act, 65; blocks environmental cleanup, 238n9

enemies, 18, 27, 31; Carl Schmidt and, 15; immunology and, 83; reforestation and, 202–4

ensembles: multispecies, 195; self and, 34–35

entanglements, 5–6; symbiotic, 4

entrepreneurs, *interessement* and, 25, 27, 33

environment: apocalyptic thinking and, 6; change and, 2–3, 227n5; destruction of, 2, 6; injustice and, 181–84, 220; management of, 17, 171–75, 177

eradication: Gambian pouched rat (*Cricetomys gambianus*), 151, 257n52; impossibility of, 129, 151, 219

escape, 220; cosmopolitical worlds and, 18; happiness and, 67, 136, 153; marketplace and, 65; reciprocal capture and, 5, 61, 68

ethnoprimatology, 106

ethology: ethograms, 115; methods, 6, 106; ultimate (evolutionary) explanation and, 6

euthanasia, 60–61, 68

exceptionalism: Donna Haraway and, 268n6; human, 18–19, 72, 84

excess, 18, 41

exotic animals, 65, 113, 126, 131–32, 134–60; in Florida, 104, 143, 151, 256n32

Exotic Pet Amnesty Day, 151

exotic species used in reforestation, 197–202

exotic tourism, 8

exotic research destinations, 10–11

expatriates, U.S.: in Panama (Zonians), 12–13

expropriation, of land, 170–71, 178–79

extinction, 4, 229n23; failure to prevent, 66; golden toad (*Incilius periglenes*), 192–94; mass, of frogs, 58, 98; Pleistocene, 186; preventing, 52

extrafloral nectaries, 32–33, 195, 203, 263n18; pictured, 24

face, 20, 234n23; monstrous faces, 59. *See also* interface

fangueo tractors, 173, 217; conservation uses, 173–75; pictured, 174, 188

farming. *See* agriculture; campesinos

figures, 7, 239n32; living, 44–46, 66; messianic, 44, 216

fingereyes (Hayward), 20–21

fire, 169, 173, 186

flagship species: golden frogs (*Atelopus zeteki*), 57; golden toads (*Incilius peri-*

glenes), 192–94; resplendent quetzals (*Pharomachrus mocinno*), 191, 206, 212

flexible accumulation, 135, 142. *See also* capitalism

flexible personhood, 135–36, 157; birds' capacity, 142; limits to, 138, 143–44; snakes' capacity, 149

Florida: Daytona Beach National Reptile Breeders' Expo, 144; exotic animals in, 104, 143, 151, 256n32; as inner frontier, 107, 110; laboratory animals, 101–3; lynchings in, 113; monkeys in, 104–33; Silver River, 110–33; State Legislature, 126–27; Trayvon Martin, 107; Wild Things zoo, 108–10. *See also* birds; snakes

flourishing, 4, 6, 217–18; blasted land-scapes and, 37, 216, 217; emergent ecologies and, 17, 36–37, 101, 153, 167; endangered species, 46, 52, 60, 156; ontological amphibians, 30, 86, 97–98, 142, 189

forest: bombing ranges, 37–39; clear cut, 1, 10, 154; cloud forests, 190–216; ecological units and, 2; fragments, 29; herbivory and, 130–31; *How Forests Think* (Kohn), 5; market economies and, 50–51, 192; Monteverde Cloud Forest Reserve, 159–61; Nathan Bedford Forest, 113; Ocala National Forest, 127–29; tropical dry forest, 167–71, 179, 186–88; tropical rain forest, 4, 8–14, 47–48, 80; understory plants, 32. *See also* deforestation; rangers; reforestation

Foucault, Michel: biopolitics, 64; *entstehung* (emergence), 227n3; gov-ernmentality, 239n24; heterotopia, 61; panopticon effect, 12

Fox television, 130

Franklin, Sarah: kinship, 256n41; wild-ness, new forms of, 67

frogs, 4, 52–71; Australian green tree frogs (*Litoria caerulea*), 153; communi-cation, 36–37; edible, 50; endangered, 55, 239n30; fringe-toed foam frogs

(*Leptodactylus melanonotus*), 163–65, 167, 188–89; Frog Forum, 152–53, 159–60; killed by chytrids, 78–79; orphans of ecosystems, 218, 258n59; pain experiences of, 95; poison dart frogs (*Dendrobates* species), 51, 156, 258n68; red-eyed tree frogs (*Agalych-nis callidryas*), 51, 100; social norms of, 153; Treewalkers International, 155–59; túngara frogs (*Engystomops pustulosus*), 36–37; *umwelt* of, 95. *See also* endangered species; *Xenopus laevis*

fruit flies. *See* mutant fruit flies

fungi, 1, 3; beneficial, 71; crop pests, 184; diversity and, 4; domestication and, 5; filamentous, 83; pathogenic, 4, 61; "thinking like a fungus," 5; zoosporic, 74–75. *See also* chytrid fungi

future, 2–3, 6–8, 11; hope and, 43, 58; speculation and, 70; uncertainty and, 13

gender, 81; postapocalyptic possibilities, 215–16

genetics: altruism and, 28; deformity and, 140–42; experimentation by animals, 98; genetic capital, 139–40; genetic determinism, 15; indetermi-nate, 82; polyploidy, 248n41; value and, 60, 64–65, 158

Gleason, Henry: Clements/Gleason debate, 2, 227n4; contingency of ecological communities, 2; theories proved, 3

global warming. *See* climate change

golden frogs (*Atelopus zeteki*), 51, 57, 60–66, 242n24; killed by conserva-tionists, 60, 243n25

Gore, Al: *An Inconvenient Truth*, 58

gringos, 39–40; refusing to leave, 42; as sticky objects, 40

Gruen, Lori: entangled empathy, 34

guardabosques (forest rangers), 10, 13, 17, 174

Guatemala, 24–25, 32

(*Hyparrhenia rufa*), 169–71, 186; love and, 218; naturalized, 130; situated judgments and, 202–4; uncertainty and, 264n30; unproblematic nature of, 151; xenophobia and, 100–101, 104, 130, 218. *See also Xenopus laevis*
IUCN Red List, 158, 213

James, William: selfhood, 34
Jameson, Fredric: messianic thought, 238n16, 239n17
Janzen-Connell hypothesis, 4
jeremiad narratives, 58

Kant, Immanuel, 231n23
killing: necessity of, 218–19, 267n6; predators eliminated from ecosystems, 176; "Thou shalt not make killable" (Haraway), 249n54, 257n52. *See also* conservation; death
kin recognition, 15–16
kinship, 18, 137–38; ethnocentric assumptions and, 138; interspecies, 136, 254n12; queer, 135–36, 149; technology, 256n41
Kockelman, Paul: ensembles of selves, 34; parasites, 228n22
Kohn, Eduardo, 5; selfhood, 34

labor: animals and, 246n31, 254n5; care and, 5, 52–57, 61–62, 153–61; ecological, 84; emotional, 56, 135; imaginative, 6, 56–59, 69, 194; producing excess, 28–29; race and, 11; wage, 13
landscapes: fragmented, 1, 206; three-dimensional illusions and, 8. *See also* blasted landscapes
Latour, Bruno: speech prosthetics, 165–66, 208; factishes, 234n29; oligopticon, 243n35
LEED gold certification for environmental design, 43
Leibniz, Gottfried: on love, 44–46
Lévi-Strauss, Claude, 43
lines of flight, nomadic animals and, 14

litter, 48; leaf, 14; leaf and plastic, 29, 210; reforestation uses, 198
lively capital: defined, 134–36; endangered species and, 154–55; pet breeding and, 141, 149–50
loneliness, artificial ecosystems and, 62
Longcore, Joyce, 73–84
love, 3, 74, 218–20; capitalism and, 135–36; death and, 142; extinction and, 218; interspecies, 46, 61, 135–36, 149; politics of, 218; unloved species, 46, 52, 57, 72
lynching, 113, 251n34. *See also* race

machines, conservation and, 173–75, 180
Macondo (McOndo), 40
Marder, Michael, 5, 237n73
Márquez, Gabriel García: *One Hundred Years of Solitude*, 40
Martin, Trayvon, 107
Marx, Karl, 134; capital as nonhuman agent, 168
Mascaro, Joseph, 2, 227n5
material-semiotic exchanges, 27, 31; trophallaxis and, 28, 243n42
Matsutake Worlds Research Group, 5, 106
Maya, Q'eqchi': ant-plant symbiosis and, 24
messianic thinking, 40, 57–58; biology and, 70, 242n15, 244n60; rejected, 215–16. *See also* Derrida, Jacques
metamorphosis, 78
methods, 6, 61–62; appropriating experimental apparatuses, 18–19, 28, 30–31, 236n46; art and, 62–65; behavioral ecology, 14–15, 29; DNA testing, 91, 93–94, 98–99; ethnoprimatology, 106; para-ethnographic, 65; performative experiments, 85, 91–96; self-experimentation, 94; visual anthropology, 106. *See also* ethology; microscopy; multispecies ethnography
microbes, 5; awakening pleasures, 219; cultures, pictured, 73–74. *See also* bacteria; fungi

microbiome, 69–70

microscopy, 6, 14, 76, 79; sensory prosthetics, 73–74

modernity: excess of world system, 29; magical realism and, 40

monkeys: howler, 205; risk of contact with, 110; sighted in Florida, 104; Togean macaques, 131; tourists attracted by, 195. *See also* Mystery Monkey of Tampa Bay; rhesus macaques

Monteverde Cloud Forest School, 194–216; pictured, 196, 199, 205

More, Sir Thomas: *Utopia*, 12

multispecies assemblages, 3

multispecies communities, 3–5

multispecies ethnography, 5–6

multispecies families, 135–62

multispecies migrations, 1, 3

multispecies ontologies, 229n31

Multispecies Salon, The, 64, 91, 243n36

multispecies worlds, 5–6, 63

mutant fruit flies (*Drosophila* species), 47, 53, 63, 67; emergent ecologies and, 243n32

Mystery Monkey of Tampa Bay, 105–6, 108–10

NASA, 48–49; frogs in space, 97

National Rifle Association (NRA), 111

nature: democracy and, 166; as infrastructure, 12; natural selection, 16; reified, 8, 10; speaking for, 165, 175, 257n52

Nature Conservancy: Department of Defense and, 37, 238n9; predatory lending by, 197

nerve gas, 37

networks, 16. *See also* actor-network theory

New York City. *See* Brooklyn

nomads, 13–16; capitalism and, 168–69; frogs as, 86, 103–4; irredeemably destructive (Stengers), 18, 101, 213n23; migrant labor, 185; social worlds and, 27–28

Noriega, Manuel, 13

nostalgia, for U.S. empire, 48

novel ecosystems, 2–5, 133, 151, 175; compost and, 201. *See also* emergence

Obama, Barack: as messianic figure, 42

oblique powers (García Canclini), 57, 161, 166, 189, 211

ontological amphibians, 4, 72; choreographing human ontologies, 97; defined, 18; disrupting human dreams, 171; flexible personhood and, 142; flourishing with capitalism, 30, 136, 189; frogs as, 37, 69, 96–98; insects as, 18–21, 31–32, 229n24; limits of world, 32; literal amphibians, 44; microscopic fungi as, 72–78; multispecies families and, 158–59

ontological cage, 19, 32, 73

ontological choreography, 84, 96–98, 246n33

organic intellectuals, 44, 194, 219

Ortiz, Fernando: plants as agents of transculturation, 185–86, 228n11

pain: frogs' experience of, 94–95

paleontology, 3, 214

Panama: Canal Treaty (1977), 40; El Giral, 29–31; El Valle, 52–54, 68; founding of the nation (1903), 9; Gamboa, 11, 13, 14; Guna Yala, 36; San Antonio, 47–51; segregation, 11; sovereignty transferred (1999), 13, 40–41; Tropic Survival School, 48–49; U.S. invasion of (1989), 13, 42; U.S. occupation of (1903–99), 9–13, 40–41. *See also* Canal Zone; City of Knowledge; Clayton Army Base; Reverted Zone

Panama Canal, 9–12; environmental impacts of, 10; French failure and, 10; pictured, 30

para-ethnographic objects, 65

parallax effect, 8

parasites, 17–18, 32, 182–85; agroindustrial ecosystems and, 13, 182–85; capitalism and, 168, 184; endangered frogs and, 55, 61; generating diversity,

4, 201; gift of hospitality, 100; host interactions, 4; landlords as, 182; Michel Serres and, 4, 100; paraselves and, 82; Paul Kockelman and, 228n22; poaching and, 232n3; resistance to, 86. *See also* chytrid fungi; worms

Pareto, Vilfredo, 28–29

parrots (order: Psittaciformes): Australian budgies (*Melopsittacus undulatus*), 138; bourkes (*Neopsephotus* species), 139, 143; Carolina parakeet (*Conuropsis carolinensis*), 143; cockatiels (*Nymphicus* species), 140–41, 143; Florida's introduced species, 143, 151, 256n32; golden conures (*Aratinga solstitialis*), 141, 143; Hahn's macaws (*Diopsittacca nobilis*), 137–38; Sunday conures (*Aratinga nenday* x *Aratinga solstitialis*), 141

past: conservation and, 166, 173, 214–15; evolution and, 11; nostalgia, 174

pathogens: diversity and, 4; endangered frogs and, 44

People for the Ethical Treatment of Animals (PETA), 92–94; Audubon Society fights against, 126

performative experiments, 85, 91–94, 96

personhood: commodified animal persons, 137; flexible persons, 135–36; snakes as lacking, 149

pesticides, 184–85

pet trade: abandoned animals, 109; animals escape from, 104, 136, 142–43; bird, 137–43; endangered species and, 154–60; extinction and, 55, 65–66; frog, 152–61; illegal, 134–36, 254n2; salvation and, 154; snake, 143–51; spreading disease, 90–94

pharmakon, 39; hope and the, 42, 66

Planned Parenthood, 88; Margaret Sanger, 90

plants: animal rhizomes, 169, 186; ant associations, 23–24, 30, 31–34; awakening pleasure with fruit, 195, 204–5; cattails (*Typha domingensis*), 171–75, 179–81, 188; *Cecropia* trees, 30; cli-

mate change and, 212–14; convivial, 5, 195–97; cultural change agents, 228n11; diversity and, 4, 195–98, 200, 266n57; egoistic, 202–3; emergent, 1, 203–4, 210–11; endangered, 213; epiphytes, 197; ice cream bean trees (*Inga* species), 24, 32, 195–96, 210; *interessement* and, 195–96, 203; monkey boogers (*Saurauia montana*), 205; *ojo de buey* liana (*Mucuna pruiens*), 48; as ontological amphibians, 4; palo verde trees, 167, 173; pollination, 32–33; *Pseudobombax* trees, 29; "question of the plant," 5; rhizomes, 169, 172–73; as subaltern, 237n73; thinking, 5; useful, 205, 208–9; water hyacinth, 167, 171; wild avocado family (Lauraceae), 206–8, 211–12. *See also* cotton; invasive species; Marder, Michael; rhizosphere; rice; seeds; sugar

poaching, 10–13, 171, 181; reading as, 232n3

politeness: human norms taught to animals, 137; love and, 220; respecting animal norms, 153; tactful distance and plants, 209–10, 216

political economy: agriculture and, 168–70; classification and, 80; dreamworlds and, 43–44; extermination and, 151; frog conservation, 66; love and, 218; knowledge and, 10; pesticides and, 218; pet breeding and, 135; spectacular nature and, 38, 167, 175–81, 191; tactical biopolitics and, 161

pollution, endangered species flourishing in, 161

posthuman utopia, 35, 39, 98

precarity: capitalism and, 150, 157, 184; global environments and, 192; indigenous adaptations and, 214–15; parasites and, 184; precarious life, 55; *precarista* movement, 178–79; shared vulnerability, 126; unequal vulnerability, 107, 120–29, 192

predator-prey entanglements, 4, 153

pregnancy tests: frogs used in, 88–93; mice killed for, 89; ontological choreography and, 96–97; pee-stick test, 96; rabbits killed for, 89

queer kinship, 135–36, 149

race: lynchings in Florida, 113; segregation in Florida, 112–14, 132; segregation in Panama, 11; whiteness, 11
rangers (*guardabosques*), 10, 13, 17, 174
rationality, 18, 28, 31; pleasures eclipse, 219–20
reciprocal capture (Stengers), 3–4, 24, 74; escape from, 5, 18, 35, 220; phenomenological worlds and, 23; symbiotic agreement and, 4, 159
reciprocity, 15, 18, 34
reforestation, 32, 186, 194–216; laissez-faire policies and, 209. *See also* forest
reproduction: birth control, 89; clones, 80–81; inbreeding, 140–41; pregnancy testing, 88–97; sexual selection, 165. *See also* sex
Reverted Zone (Panama), 17–18, 35–51; biosecure amphibian pods, pictured, 45; building, pictured, 41; Wounaan village, pictured, 50. *See also* Canal Zone; City of Knowledge
rhesus macaques (*Macaca mulatta*), 105–33; ban on trapping, 130, 132; bites by, 125; breeding season of, 252n55; diseases of, 109–10; distribution within Florida, 129; dominance hierarchies of, 115–19; flourishing amid extinction, 218; global range of, 107; introduced to Silver River (Florida), 111; as invasive species, 130; mourning by, 129; observing human behavior, 114; social behavior, 114–16; trapping of, 127–29
rhizomes, 169, 172–73; cattails, pictured, 172; resisting capital, 180; stolons compared with, 203; sugar cane, 185–86
rhizosphere, 198–202

rice (*Oryza sativa*), 179–88; *arrozon* parasites (*Oryza latifolia*), 182
risk, and wildness, 67, 107, 112, 152
rivets: added to agro-ecological systems, 160; bombs as, 39; cattle as living, 186; ecosystem, 2–3, 188, 192, 218; machines as, 173, 180; plants as novel, 214
Roosevelt, Teddy, 9, 25
Rose, Deborah Bird: animal faces, 20, 234n23; care, 34; death, 202; ethics of killing, 61, 69, 202; wildness, 106
ruins: Bettina Stoetzer on, 267n62; gardening in, 214–16, 219

San Francisco, Golden Gate Park, 100–101
Sanger, Margaret, 90
scarcity: economics of, 28–29, 41; of water, 12
Schmitt, Carl: enemy-ally distinction, 15, 83, 101
Schneider, David: kinship, 137–38
Science Wars, 28
seeds: animal rhizomes, 169–70; caring for, 198; dispersed by birds, 33, 203–4; generating multispecies becomings, 195–96; orphaned by extinction, 186–87; wind dispersal, 173
segregation: ecological science and, 11, 13, 166, 179; Florida and, 111–14, 132; U.S. policy in Panama, 11–12
selfhood, 34–35; paraselves, 82
selfish genes, 16
Serres, Michel: *The Parasite*, 4, 100, 228n22
sex: interspecies kink, 138–39; parasexuality, 81. *See also* reproduction
Sloterdijk, Peter: ontological amphibians, 18–20, 229n24; spheres, 76, 82–84
Smithsonian Tropical Research Institute, 8–11, 13–14, 17; Insect Cognition Laboratory, 17
snakes: ball pythons (*Python regius*), 147; Burmese pythons (*Python molurus bivittatus*), 104, 151–52, 219; chondros

(*Chondropython viridis* or *Morelia viridis*), 143–52; force-feeding of, 144–45
social insects: altruism and, 28; learning and, 21
social worlds, 11–12, 17; of ants, 27–28; of macaques, 117–19
sociobiology, 15
soil, 198–202; compaction by cattle, 200; enlivened by worms, 201–2; Kristina Lyons, 265n39
species: diversity of, 4; exceptional, 15, 17; human constructs, 80, 140–41; invasive, 13; known unknown, 80; nomadic, 1, 3, 13–14; novel, 141, 143; as rivets, 2–3. *See also* endangered species
spectacles, 146; conservation and, 182, 191; multispecies, 254n5
Star, Susan Leigh, 136; architectures of apartheid, 230n13; torque, 255n16
Stengers, Isabelle: cosmopolitics, 18; *pharmakon*, 39; on reciprocal capture, 3–4; on symbiotic agreements, 4, 228n16; utopia, 62, 65
sticky objects, 40, 56
subaltern, plants as, 237n73
Subway restaurant: destroyed, 43; as middle-class utopia, 41
succession, 2, 218
sugar, 185–86, 228n11
superorganisms, 34; ant colonies as, 15–16; ecosystems as, 2
surplus, 18, 29
surrogates, Pleistocene extinction and, 186
surveillance, 12, 64; by animal activists, 127
swarms: insect, 13; lively agents, 215; microbe, 82, 201; monkey, 122; politics of, 229n25
symbiosis, 8; compost and, 201; facultative, 24, 32–34; Lynn Margulis and Dorion Sagan on, 4; multispecies families and, 159; reciprocal capture and, 4, 24; symbiotic agreements, 3, 228n16. *See also* Stengers, Isabelle

Tansley, A. G., 2, 227n5
Tarzan, 109–14; pictured, 112; Tarzan Swing, 207
technology: barbed wire, 168; hunting, 176; parasites of, 182. *See also* machines; microscopy
thievery, 28, 31
thirdness (Peirce), parasites and, 228n22
Thompson, Charis (*née* Cussins): ontological choreography, 96, 246n33
tick: poverty of world, 19; senses of, 233n15
time, 10; homogeneous and empty, 40; messianic, 40, 58–59; time-place learning, 23; travel and, 8, 166. *See also* future; past
tolerance (Stengers), 18
torque (Bowker and Star), 255n16; animal personhood and, 136, 139
Torrijos, Omar, 40
tourism, 8; ecotourism, 17, 181, 198, 209; extinction and, 192–93; inequality and, 190–92, 206
toxins: impacts on human bodies, 185; pesticides, 184–85, 189; plant, 173. *See also pharmakon*
trees. *See* plants
trophallaxis (nourishment/interchange), 28–31; defined, 236n54; material-semiotic exchanges, 235n42
tropics: time travel and, 8, 166; tropical ecology, 8, 10. *See also* forest; Smithsonian Tropical Research Institute
Tsing, Anna: contaminated diversity, 210; Matsutake Worlds Group, 5; unexpected connections, 229n34

umwelt, 19–22; bacterial, 76, 245n11; Brett Buchanan on, 234n19; environment (German), 233n9; frog, 95, 98; Giorgio Agamben on, 233n12; Merleau-Ponty on, 234n19; Peter Sloterdijk on, 73, 84; suffering capacity and, 95
unexploded ordinance (UXO), 37–39